Australian

Signpost MATHS

2

Alan McSeveny
Diane McSeveny-Foster
Erika Johnson
Alan Parker
Rachel McSeveny

Pearson Australia
(a division of Pearson Australia Group Pty Ltd)
459–471 Church Street, Level 1, Building B, Richmond, Victoria 3121
PO Box 23360, Melbourne, Victoria 8012
www.pearson.com.au

First published 2018 by Pearson Australia
2025 2024 2023 2022
10 9 8 7 6 5 4 3 2 1

Content Learning Specialist: Julian Lumb
Project Manager: Shelly Wang
Production Manager: Elizabeth Gosman & Aptara
Editor: Aptara
Designer: Anne Donald & Jennifer Johnston
Cover designer: Jennifer Johnston
Desktop operator: Aptara
Rights & Permissions Editor: Samantha Russell-Tulip
Senior Publishing Services Analyst: Rob Curulli
Cover art: Michael Barter
Cover background: rf_vector/shutterstock.com
Illustrators: Michael Barter, Wendy Gorton, Nives Porcellato, Andrew Craig
Printed in Malaysia (CTP-VVP)

ISBN 978 1 4886 2181 9
Pearson Australia Group Pty Ltd ABN 40 004 245 943

Acknowledgements
All material identified by ACARA is material subject to copyright under the *Copyright Act 1968* and is owned by the Australian Curriculum, Assessment and Reporting Authority 2018. ACARA neither endorses nor verifies the accuracy of the information provided and accepts no responsibility for incomplete or inaccurate information. In particular, ACARA does not endorse or verify that:
- The content descriptions are solely for a particular year and subject;
- All the content descriptions for that year and subject have been used; and
- The author's material aligns with the Australian Curriculum content descriptions for the relevant year and subject.

You can find the unaltered and most up to date version of this material at http://www.australiancurriculum.edu.au/ This material is reproduced with the permission of ACARA.

Every effort has been made to trace and acknowledge copyright. However, if any infringement has occurred, the publishers tender their apologies and invite the copyright holders to contact them.

What is Australian Signpost Maths?

Australian Signpost Maths is a mathematics activity book series for students from Foundation to Year 6. The series has been written to meet the requirements of the Australian Curriculum.

The components of the series include Student Books, Teacher's Books, Mentals Books and an interactive Website. Teachers can select an appropriate program for every student from the rich and varied material provided.

The content has been carefully sequenced within each year level and across the series to take into account students' likely mathematical development.

Student Books

Teacher's Books

Mentals Books

Website

Structure of Australian Signpost Maths

Australian Signpost Maths emphasises the curriculum's syllabus content as well as problem-solving strategies, language development and the use of technology.

To maximise the benefits of the program, the Student Book, Teacher's Book, Mentals Book and Website should be used together.

The sequence of units in the **Student Book** forms a suggested program for the year. The **Teacher's Book** also provides lesson plans for each page of the **Student Book**, and blackline masters to assist teachers in implementing the program.

The Student Book presents lessons as a mix of content strands. However, the Contents and Contents Cross-reference pages in the Student Book allow teachers to construct programs based on the specific content strands (Number and Algebra, Measurement and Geometry, and Statistics and Probability). Progress Tests and remediation records are located in the Teacher's Book and on the website. These tests are also now included in the back of this book.

The **Mentals Book** mixes examples from all content strands, reviewing the content of previous units of the Student Book.

The innovative **Website** help teachers to bring mathematics alive with technology. The website provides interactive maths tools, games and practice opportunities as well as relevant resource masters and worksheets for all year levels. These can be used for whole-class, small-group and individual learning. The website also includes **Concept Check-In**, a new diagnostic screener.

Special Features of Australian Signpost Maths

- **Traffic Light** system allows students to reflect on their work and highlight any units that they are having trouble understanding. They tick the red for units they feel they still don't understand, and green for those they feel they understand fully.

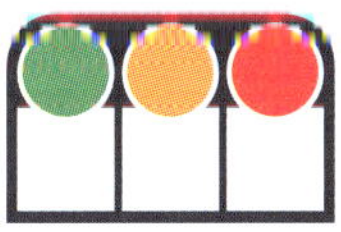

- Exercises are **well graded**. Work is reinforced in the Mentals Book.
- The **Progress Tests** (now also in the back of this book) allow the teacher to discover each student's strengths and weaknesses, and the cross-references direct students to the pages where that work is introduced.
- **Answers** are supplied in the Teacher's Book.
- The **Dictionary** at the beginning of this Student Book will help students to learn the language of mathematics.
- **ID Cards** (in the Mentals Book, Teacher's Book and Website) review the language of mathematics by asking students to identify common terms, shapes and symbols.
- Important **rules and concepts** are clearly highlighted.
- **Worked examples** and explanations are given throughout the Student Book where new ideas are introduced.
- The use of **colour** makes emphasis clear and is highly motivating.
- **Cartoons** give instruction and friendly advice.
- **Interactive Activities** are provided on the website for whole-class, small-group and individual learning.

Australian Signpost Icons

Signpost icons are used throughout the book as cues to the essential nature of exercises and activities, and as a guide to ways of engaging with them. These icons often indicate alternative or more concrete approaches to dealing with concepts.

This icon highlights **important rules and concepts** occurring throughout the book. It often appears with worked examples.

Activities provide **applications and enrichment**. These activities usually involve the use of concrete materials and partner or group work.

These enjoyable activities are used to **motivate and involve** students in mathematical pursuits. They usually involve games and puzzles.

Investigations allow students to **explore and discover** maths concepts.

This icon indicates the use of computers, calculators or other **information and communications technology**.

Australian Curriculum Proficiency Strands

The proficiency strands of the Australian Curriculum describe how content is explored or developed – that is, the 'thinking and doing' of mathematics.

Understanding

Learning the concepts

*Students build a robust knowledge of adaptable and transferable mathematical concepts. They make connections between related concepts and progressively apply the familiar to develop new ideas. They develop an understanding of the relationship between the 'why' and the 'how' of mathematics.**

Conceptual understanding of maths ideas includes the explanation of a concept using text and diagrams. This occurs throughout Australian Signpost Maths at the top of many pages and is indicated by the Concept icon.

Fluency

Using the concepts

*Students develop skills in choosing appropriate procedures, carrying out procedures flexibly, accurately, efficiently and appropriately, and recalling factual knowledge and concepts readily.**

The practice of maths skills to build fluency occurs on every page of Australian Signpost Maths.

*The Australian Curriculum: Mathematics, v1.2 – Content structure

Problem Solving

Applying concepts and strategies to develop solutions to problems

*Students develop the ability to make choices, interpret, formulate, model and investigate problem situations, and communicate solutions effectively.**

Problem solving provides opportunities for students to use strategies and skills such as investigating and questioning, to collaborate with others and to communicate their findings to different audiences. Such activities are often indicated throughout Australian Signpost Maths by the Activity and Investigation icons.

Reasoning

Coherent and logical thought

*Students develop an increasingly sophisticated capacity for logical thought and actions, such as analysing, proving, evaluating, explaining, inferring, justifying and generalising.**

Students require opportunities to explain their mathematical thinking and can do so through both diagrams and written explanations. Reasoning questions are located throughout Australian Signpost Maths.

2 Contents and Syllabus Overview

KEY

- Number & Algebra
- Measurement & Geometry
- Statistics & Probability

Page	Unit	Title	Strand: Number and Algebra	Strand: Measurement and Geometry	Strand: Statistics and Probability	Sub-strand: Number and place value	Fractions and decimals	Money and financial mathematics	Patterns and algebra	Using units of measurement	Shape	Location and transformation	Chance	Data representation and interpretation	Suggested progress
1		Thinking Skills													Term 1
2	1A	Addition Combinations to 10	■			●									
3	1B	Subtraction to 10	■			●									
4	1C	One Half	■				●								
5	1D	Parallel Lines		■							●				
6	2A	Ordinal Numbers	■			●				●					
7	2B	Position Words		■								●			
8	2C	Revision of Time		■						●					
9	2D	Ordering Capacities		■						●					
10	3A	Counting Patterns	■			●			●						
11	3B	Counting	■			●			●						
12	3C	One Quarter	■				●								
13	3D	Describing 3D Objects		■							●				
14	4A	Addition to 20	■			●									
15	4B	Addition to 20	■			●									
16	4C	Comparing Masses		■						●					
17	4D	Balance Scales		■						●					
18	5A	Modelling Numbers	■			●									
19	5B	Doubling and Near Doubling	■			●									
20	5C	2D Shapes		■							●				
21	5D	Informal Units of Length		■						●					
22	6A	Groups and Rows	■			●									
23	6B	Groups and Rows	■			●									
24	6C	Analogue Time		■						●					
25	6D	Analogue Time		■						●					T1*
26	7A	Sharing	■			●									
27	7B	Sharing	■			●									
28	7C	Graphing the Weather			■									●	
29	7D	Thinking About Graphs			■									●	
30	8A	Subtraction to 20	■			●									
31	8B	Addition to 20	■			●									
32	8C	Half of a Group	■				●								
33	8D	Quarter of a Group	■				●								

Page	Unit	Title	Strand	Number and Algebra	Measurement and Geometry	Statistics and Probability	Sub-strand	Number and place value	Fractions and decimals	Money and financial Mathematics	Patterns and algebra	Using units of measurement	Shape	Location and transformation	Chance	Data representation and interpretation	Suggested progress
34	9A	Numbers to 150		■				●									Term 2
35	9B	Numbers to 1000		■				●									
36	9C	Features of 2D Shapes			■								●				
37	9D	Looking at 3D Objects			■								●				
38	10A	Problem Solving		■				●									
39	10B	Multiplication Sign		■				●									
40	10C	Chance				■									●		
41	10D	Area			■							●					
42	11A	Subtraction to 20		■				●									
43	11B	Number Lines		■				●									
44	11C	Number Lines		■				●									
45	11D	Cones and Cylinders			■								●				
46	12A	Fair Shares		■				●									
47	12B	Division		■				●									
48	12C	Halves and Quarters		■					●								
49	12D	Halves and Quarters		■					●								
50	13A	Linking Addition and Subtraction		■				●									
51	13B	Linking Addition and Subtraction		■				●									
52	13C	Left and Right			■									●		●	
53	13D	The Trapezium and Kite			■								●				
54	14A	Sharing		■				●									
55	14B	Using Groups		■				●									
56	14C	Digital Time			■							●					
57	14D	Lists, Graphs and Tables				■										●	
58	15A	Numbers to 1000		■				●									
59	15B	Numbers to 1000		■				●									T2*
60	15C	Reading Graphs				■										●	
61	15D	Using Graphs				■										●	
62	16A	Ordering Numbers		■				●			●						
63	16B	Addition by Looking for Tens		■				●									
64	16C	Fraction of a Whole		■					●								
65	16D	Fraction of a Group		■					●								
66	17A	Adding 10s		■				●			●						
67	17B	Adding and Subtracting 10s		■				●									
68	17C	Analogue Time			■							●					
69	17D	Months of the Year			■							●					
70	18A	How Many More?		■				●									
71	18B	Flipping a Shape			■									●			
72	18C	Chance				■									●		
73	18D	Sliding a Shape			■									●			

Page	Unit	Title	Strand: Number and Algebra	Strand: Measurement and Geometry	Strand: Statistics and Probability	Sub-strand: Number and place value	Sub-strand: Fractions and decimals	Sub-strand: Money and financial mathematics	Sub-strand: Patterns and algebra	Sub-strand: Using units of measurement	Sub-strand: Shape	Sub-strand: Location and transformation	Sub-strand: Chance	Sub-strand: Data representation and interpretation	Suggested progress
74	19A	Numbers to 1000	■			●									
75	19B	Number Patterns	■			●			●						
76	19C	Balance Scales		■						●					Term 3
77	19D	Ordering Masses		■						●					
78	20A	Value of Coins	■			●		●							
79	20B	Value of Coins	■			●		●							
80	20C	Comparison of Areas		■						●					
81	20D	Area Using Informal Units		■						●					
82	21A	Multiplication Sign	■			●									
83	21B	Multiplication Sign	■			●									
84	21C	The Parallelogram		■							●				
85	21D	Using Tally Marks			■									●	
86	22A	Building to the Next 10	■			●									
87	22B	Building to the Next 10	■			●									
88	22C	Pyramids and Prisms		■							●				
89	22D	Recognising 3D Objects		■							●				
90	23A	Australian Coins	■			●		●							
91	23B	Money	■			●		●							
92	23C	Making Graphs			■									●	
93	23D	Tally Marks			■									●	
94	24A	Number Patterns	■			●			●						
95	24B	Amounts to $2	■			●		●							
96	24C	Fractions	■				●								
97	24D	Giving Directions		■								●			T3*
98	25A	Multiplication Problems	■			●									
99	25B	Multiplication	■			●									
100	25C	Rounding	■			●									
101	25D	Telling the Story from Data			■									●	
102	26A	Split Strategy (Addition)	■			●									
103	26B	Split Strategy (Subtraction)	■			●									
104	26C	Seasons		■						●					
105	26D	Months and Seasons		■						●					
106	27A	Shopping	■			●		●							
107	27B	Chance			■								●		
108	27C	3D Objects		■							●				
109	27D	Drawing 2D Shapes		■							●				
110	28A	Division	■			●									
111	28B	Repeated Subtraction	■			●									
112	28C	The Calendar		■						●					
113	28D	The Calendar		■						●					
114	29A	Jump Strategy (Addition)	■			●									

Page	Unit	Title	Strand	Number and Algebra	Measurement and Geometry	Statistics and Probability	Sub-strand	Number and place value	Fractions and decimals	Money and financial mathematics	Patterns and algebra	Using units of measurement	Shape	Location and transformation	Chance	Data representation and interpretation	Suggested progress
115	29B	Jump Strategy (Subtraction)		■				●									
116	29C	Dice, Cards and Spinners				■									●		
117	29D	Following Instructions			■									●			
118	30A	The Division Sign		■				●									Term 4
119	30B	Division as Repeated Subtraction		■				●									
120	30C	Comparing Areas			■							●					
121	30D	Gathering Data				■										●	
122	31A	Multiplication		■				●									
123	31B	Division as Repeated Subtraction		■				●									
124	31C	Turning a Shape			■									●			
125	31D	Turning Shapes			■									●			
126	32A	Subtraction Using Blocks		■				●									
127	32B	Addition Using Blocks		■				●									
128	32C	Quarters and Eighths		■					●								
129	32D	Naming Fractions		■					●								
130	33A	Choosing a Strategy		■				●									T4*
131	33B	Choosing a Strategy		■				●									
132	33C	Faces, Edges and Corners			■								●				
133	33D	Comparing Volume			■							●					
134	34A	Problem Solving		■				●									
135	34B	Counting Coins		■						●							
136	34C	Informal Units of Length			■							●					
137	34D	Comparing Objects			■							●					
138	35A	Using a Calculator		■				●									
139	35B	The Calculator		■				●			●						
140	35C	Flip, Slide and Turn			■									●			
141		Appendix 1: Australian Money															
142		Appendix 2: Money															

*Suggested placement for Progress Tests 1 to 4 (see the Teacher's Book). It is assumed that there are 10 weeks in each term.

Contents Cross-reference

Number and Algebra

1 Counting	Pages	Australian Curriculum Reference
Counting to and from any starting point	10, 11, 18, 59	Investigate number sequences, initially those increasing and decreasing by twos, threes, fives and tens from any starting point, then moving to other sequences (ACMNA026); Group, partition and rearrange collections up to 1 000 in hundreds, tens and ones to facilitate more efficient counting (ACMNA028)
Number sequences of twos, threes, fives and tens	10, 11, 18, 75, 79, 94, 139	Investigate number sequences, initially those increasing and decreasing by twos, threes, fives and tens from any starting point, then moving to other sequences (ACMNA026)
Using a calculator	10, 11, 102, 103, 126, 127, 130, 138, 139	Investigate number sequences, initially those increasing and decreasing by twos, threes, fives and tens from any starting point, then moving to other sequences (ACMNA026)
Ordinal numbers	6, 112, 113	Recognise, model, represent and order numbers to at least 1 000 (ACMNA027)
2 Numeration		
Numbers to 1 000	10, 11, 18, 34, 35, 58, 59, 74	Investigate number sequences, initially those increasing and decreasing by twos, threes, fives and tens from any starting point, then moving to other sequences (ACMNA026); Recognise, model, represent and order numbers to at least 1 000 (ACMNA027); Group, partition and rearrange collections up to 1 000 in hundreds, tens and ones to facilitate more efficient counting (ACMNA028)
3 Place value		
Grouping in tens and in hundreds	34, 35, 58, 59, 74	Recognise, model, represent and order numbers to at least 1 000 (ACMNA027); Group, partition and rearrange collections up to 1 000 in hundreds, tens and ones to facilitate more efficient counting (ACMNA028)
Partitioning and regrouping numbers	58, 59, 74	Recognise, model, represent and order numbers to at least 1 000 (ACMNA027); Group, partition and rearrange collections up to 1 000 in hundreds, tens and ones to facilitate more efficient counting (ACMNA028)
4 Addition and subtraction		
Addition problems	2, 14, 15, 19, 31, 43, 50, 51, 63, 66, 67, 70, 86, 87, 102, 114, 127, 130, 131	Explore the connection between addition and subtraction (ACMNA029); Solve simple addition and subtraction problems using a range of efficient mental and written strategies (ACMNA030); Describe patterns with numbers and identify missing elements (ACMNA035)
Subtraction problems	3, 30, 42, 43, 44, 50, 51, 66, 103, 115, 126, 130, 131	Explore the connection between addition and subtraction (ACMNA029); Solve simple addition and subtraction problems using a range of efficient mental and written strategies (ACMNA030)

Mental and written strategies for addition and subtraction	2, 3, 14, 15, 19, 31, 43, 44, 63, 66, 67, 70, 86, 87, 102, 103, 114, 115, 126, 127, 130, 131	Investigate number sequences, initially those increasing and decreasing by twos, threes, fives and tens from any starting point, then moving to other sequences (ACMNA026); Solve simple addition and subtraction problems using a range of efficient mental and written strategies (ACMNA030)
Connecting addition and subtraction	50, 51	Explore the connection between addition and subtraction (ACMNA029); Solve simple addition and subtraction problems using a range of efficient mental and written strategies (ACMNA030)
5 Multiplication and division		
Multiplication and repeated addition problems	22, 23, 38, 39, 44, 82, 83, 98, 99, 118, 122, 134, 138	Recognise and represent multiplication as repeated addition, groups and arrays (ACMNA031)
Division problems	26, 27, 46, 47, 54, 55, 110, 111, 118, 119, 134, 138	Recognise and represent division as grouping into equal sets and solve simple problems using these representations (ACMNA032)
Multiplication and division strategies	22, 23, 27, 28, 38, 39, 44, 46, 47, 54, 55, 82, 83, 98, 110, 111, 118, 119, 122, 123, 134	Recognise and represent multiplication as repeated addition, groups and arrays (ACMNA031); Recognise and represent division as grouping into equal sets and solve simple problems using these representations (ACMNA032)
Connecting multiplication and division	23, 54, 55, 134	Recognise and represent multiplication as repeated addition, groups and arrays (ACMNA031); Recognise and represent division as grouping into equal sets and solve simple problems using these representations (ACMNA032)
6 Fractions		
Halves, quarters and eighths of everyday shapes and collections	4, 12, 32, 33, 48, 49, 64, 65, 96, 128, 129	Recognise and interpret common uses of halves, quarters and eighths of shapes and collections (ACMNA033)
7 Money		
Australian coins and notes	2, 78, 79, 90, 91, 95, 106, 135	Count and order small collections of Australian coins and notes according to their value (ACMNA034)
8 Patterns and algebra		
Number patterns and their rules	10, 11, 62, 66, 67, 75, 94, 139	Investigate number sequences, initially those increasing and decreasing by twos, threes, fives and tens from any starting point, then moving to other sequences (ACMNA026); Describe patterns with numbers and identify missing elements (ACMNA035)
Using number sentences to solve problems	14, 15, 19, 23, 30, 31, 39, 43, 44, 50, 51, 82, 83, 98, 99, 138	Explore the connection between addition and subtraction (ACMNA029); Solve simple addition and subtraction problems using a range of efficient mental and written strategies (ACMNA030); Recognise and represent multiplication as repeated addition, groups and arrays (ACMNA031); Recognise and represent division as grouping into equal sets and solve simple problems using these representations (ACMNA032)

Measurement and Geometry

	Length	Pages	Australian Curriculum Reference
1	**Length**		
	Length with informal units	21, 136, 137	Compare and order several shapes and objects based on length, area, volume and capacity using appropriate uniform informal units (ACMMG037)
2	**Area**		
	Compare area of different shapes	41, 80, 81, 120	Compare and order several shapes and objects based on length, area, volume and capacity using appropriate uniform informal units (ACMMG037)
3	**Volume and capacity**		
	Volume and capacity using informal units	9, 133, 137	Compare and order several shapes and objects based on length, area, volume and capacity using appropriate uniform informal units (ACMMG037)
4	**Mass**		
	Compare mass using balance scales	16, 17, 76, 77, 137	Compare masses of objects using balance scales (ACMMG038)
5	**Time**		
	Time to the quarter and half hour	8, 24, 25, 56, 68	Tell time to the quarter-hour, using the language of 'past' and 'to' (ACMMG039)
	The calendar	6, 112, 113	Use a calendar to identify the date and determine the number of days in each month (ACMMG041)
	Months and seasons	69, 104, 105, 113	Name and order months and seasons (ACMMG040); Use a calendar to identify the date and determine the number of days in each month (ACMMG041)
6	**Shape**		
	2D shapes	5, 20, 36, 53, 84, 109, 140	Describe and draw two-dimensional shapes, with and without digital technologies (ACMMG042)
	3D objects	13, 37, 45, 88, 89, 108, 132	Describe the features of three-dimensional objects (ACMMG043)
	Faces, corners and edges	45, 88, 108, 132	Describe the features of three-dimensional objects (ACMMG043)
7	**Location**		
	Using simple maps	97, 117	Interpret simple maps of familiar locations and identify the relative positions of key features (ACMMG044)
	Position	7, 52, 117	Interpret simple maps of familiar locations and identify the relative positions of key features (ACMMG044)
8	**Transformations**		
	Flip, slide, turn; half and quarter turns	71, 73, 124, 125, 140	Investigate the effect of one-step slides and flips with and without digital technologies (ACMMG045); Identify and describe half and quarter turns (ACMMG046)

Statistics and Probability

1 Data representation and interpretation	Pages	Australian Curriculum Reference
Gathering and recording data	28, 52, 61, 85, 92, 93, 101, 121	Identify a question of interest based on one categorical variable, and gather data relevant to the question (ACMSP048); Collect, check and classify data (ACMSP049); Create displays of data using lists, tables and picture graphs and interpret them (ACMSP050)
Lists, tables and graphs	28, 29, 57, 60, 61, 92, 93, 101, 121	Identify a question of interest based on one categorical variable and gather data relevant to the question (ACMSP048); Collect, check and classify data (ACMSP049); Create displays of data using lists, tables and picture graphs and interpret them (ACMSP050)
2 Chance		
The language of chance	40, 72, 107, 116	Identify practical activities and everyday events that involve chance. Describe outcomes as 'likely' or 'unlikely' and identify some events as 'certain' or 'impossible' (ACMSP047)
Comparing the likelihood of events	72, 107, 116	Identify practical activities and everyday events that involve chance. Describe outcomes as 'likely' or 'unlikely' and identify some events as 'certain' or 'impossible' (ACMSP047)

abacus

An instrument used for counting and calculating.

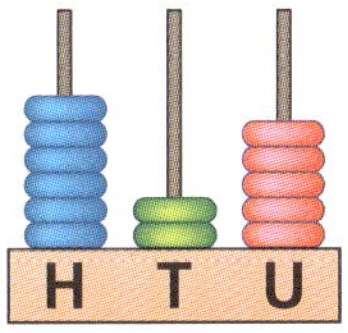

anticlockwise

A circular turn from the top to the left.

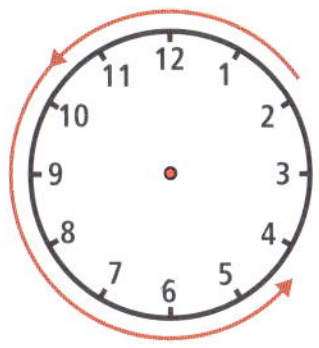

area

The amount of surface.

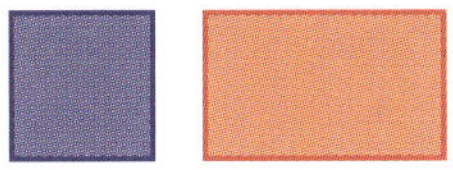

bar graph

See *graph*.

capacity

Capacity is the amount that a container can hold.

clockwise

A circular turn from the top to the right.

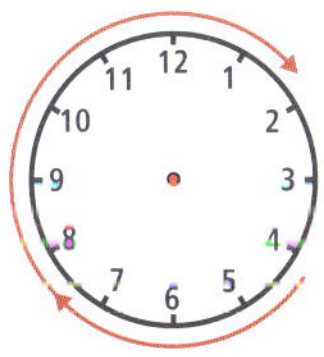

coins

5c 10c 20c 50c $1 $2

corner (vertex)

A corner is where 2 or more sides meet.

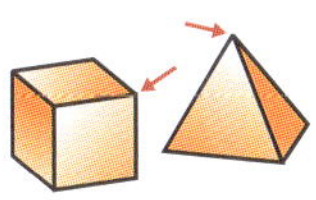

data

A collection of information.

division

Division means 'shared among', 'in groups of' or 'divided by'.

double

Double means the same thing twice.

Double 4 = 4 + 4 = 8

edge

Where two surfaces of a 3D object meet.

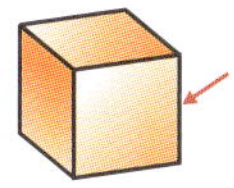

face

A flat surface of a 3D object.

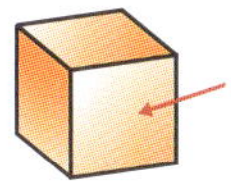

flip

A mirror image.

fraction

Any part of a whole or collection.

For example, $\frac{3}{4}$ means 3 of 4 equal parts.

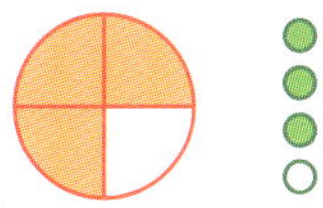

graph

A graph shows us information (or data).

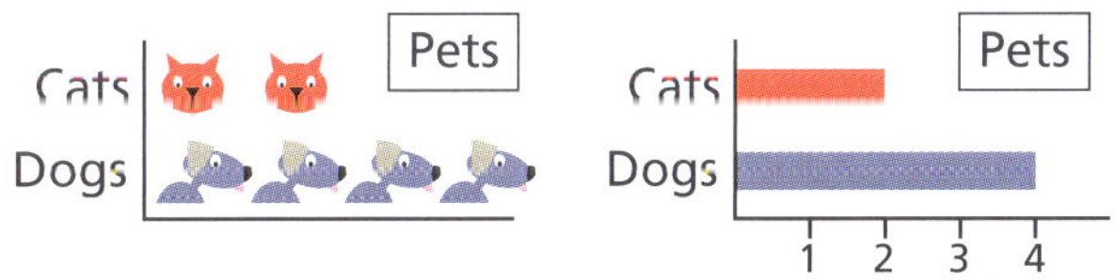

Picture Graph Bar Graph

half

One half is one of two equal parts.

This orange is cut in half.

left and right

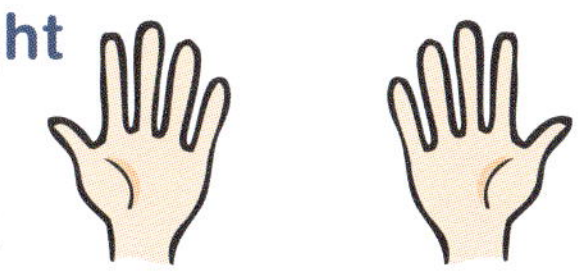

Left Right

number line

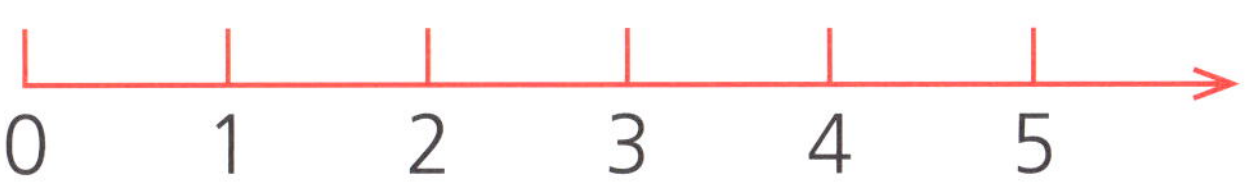

A number line is a line that shows numbers in order. Number lines can be used for counting and adding.

numeral

A written number symbol, such as 7 or 13.

ordinal number

Ordinal numbers describe the order or position of something, for example 1st, 2nd, 3rd, 4th, 15th or 31st.

partitioning

Breaking up a group of objects into parts. For example, the number 10 can be partitioned as follows:

10 = 1 + 9 = 2 + 8 = 3 + 7

picture graph

See *graph*.

place value

The position of each digit of a numeral holds a different value.

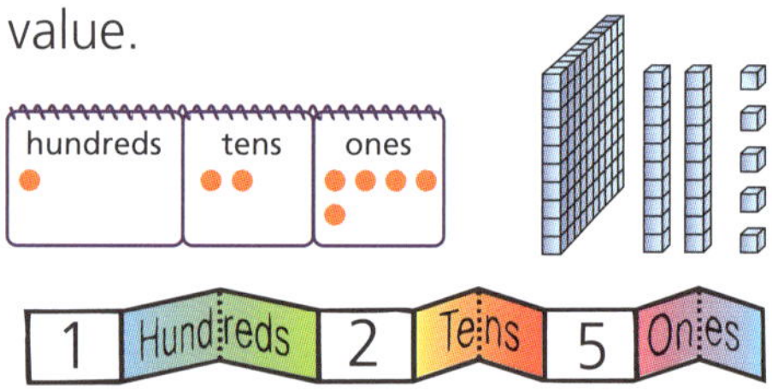

quarter

One of four equal parts.

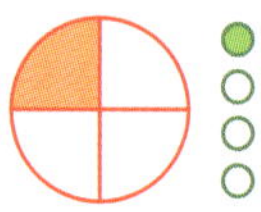

right and **left**

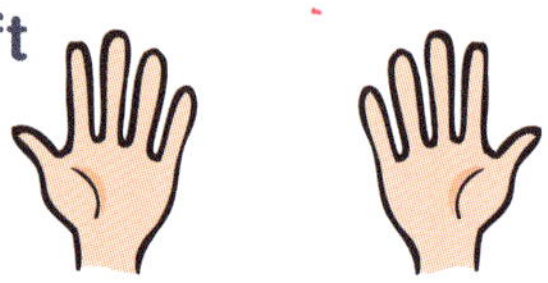

seasons

There are four traditional seasons in one year.

Summer: December, January, February

Autumn: March, April, May

Winter: June, July, August

Spring: September, October, November

sharing

When sharing, we make sure that each share is the same size.

3 people could share these 6 balls. Each person would get 2 balls.

slide

Moving a shape in any direction without changing the size or orientation.

straight line

A straight line has no bends or curves.

tally marks

Tally marks are used to keep count and show groups of 5.

 = 5

three-dimensional (3D) objects

See page xvi.

time

See page xv.

turn

Moving a shape in a clockwise or anticlockwise direction, without changing the shape or size.

two-dimensional (2D) shapes

See page xvi.

time words:

Days						
Sunday	Monday	Tuesday	Wednesday	Thursday	Friday	Saturday

Months					
January	February	March	April	May	June
July	August	September	October	November	December

- Clocks

Half past 3

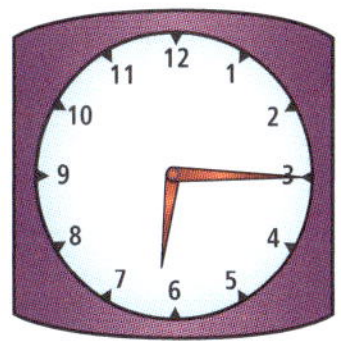

Quarter past 6

half past When the long hand is pointing to the 6 and the short hand is pointing in between two numbers, the time is a 'half past' time.

quarter past When the long hand is pointing to the 3 and the short hand is a quarter of the way past the hour number, the time is a 'quarter past' time.

Quarter to 7

quarter to When the long hand is pointing to the 9 and the short hand is a quarter of the way from the next hour number, the time is a 'quarter to' time.

two-dimensional (2D) shapes

Flat shapes are two-dimensional.
They have length and width.

Circle

1 curved side

Triangle

3 sides

Square

4 equal sides

Rectangle

2 equal long sides and 2 equal short sides, like a stretched square

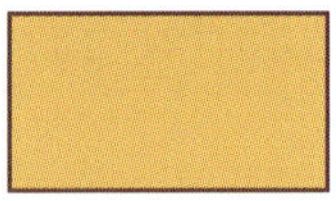

Oval

1 curved side,
like a squashed circle

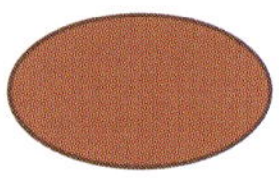

Hexagon

6 sides and 6 corners

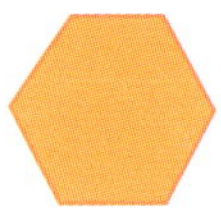

Trapezium

4 sides,
with 2 sides parallel

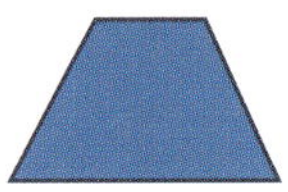

Parallelogram

2 pairs of parallel sides

Rhombus

4 equal sides, opposite sides parallel

three-dimensional (3D) objects

Solid objects are three-dimensional.
They have length, width and height.

Sphere

A sphere is curved and round.

Cube

A cube has 6 faces and straight edges.

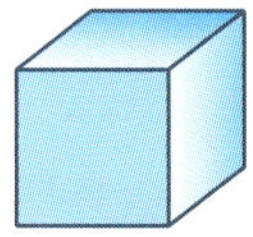

Cylinder

A cylinder has 2 faces and 1 curved surface.

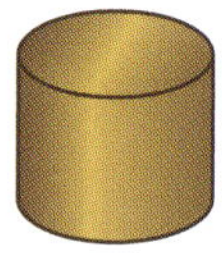

Cone

A cone has 1 face and 1 curved surface.

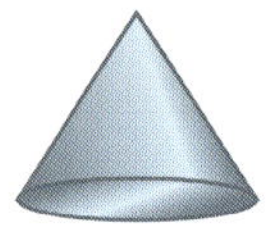

Pyramid

A pyramid has triangular faces joined around a base.

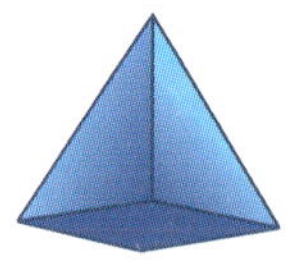

Prism

A prism has rectangular faces joining two bases.

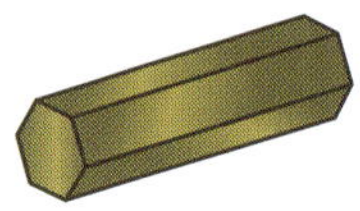

2 Thinking Skills

1. What were the mice planning to do?
2. Explain how the mice carried out their plan.
3. Why would the mice take only a part of the pie?
4. How do you think the stilts could be made?
5. Estimate the height of the stilts.
6. What could they make with the paper?
7. Does the police officer think the mice took the pie? Why or why not?
8. Make up a maths question about these pictures.

  ISBN 978 1 4886 2181 9

1A Addition Combinations to 10

Emma has

Paul has

$$\begin{array}{r} \$3 \\ +\$5 \\ \hline \$8 \end{array}$$

1 Count on from the largest number to complete these.

a 1 + 6 = ☐	**b** 2 + 2 = ☐	**c** 3 + 4 = ☐	**d** 1 + 3 = ☐	**e** 4 + 5 = ☐
f 4 + 4 = ☐	**g** 6 + 3 = ☐	**h** 5 + 4 = ☐	**i** 6 + 1 = ☐	**j** 0 + 5 = ☐
k 2 + 7 = ☐	**l** 5 + 1 = ☐	**m** 2 + 5 = ☐	**n** 3 + 6 = ☐	**o** 7 + 0 = ☐

2 Finish the **number story** for each question below.

a 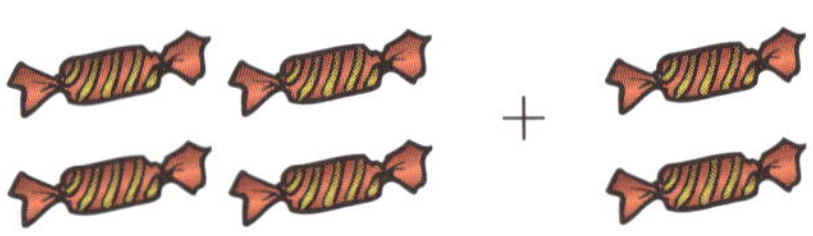

☐ lollies + ☐ lollies equals ☐ lollies.

b 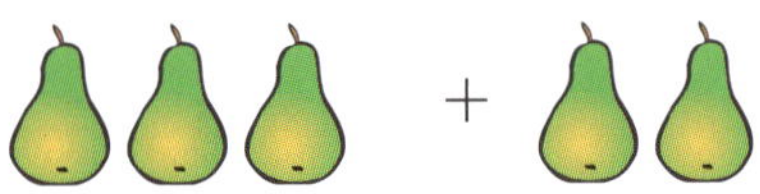

☐ pears + ☐ pears equals ☐ pears.

Use objects to model your own addition stories.

 • *AUSTRALIAN SIGNPOST MATHS 2* • ISBN 978 1 4886 2181 9

1B Subtraction to 10

1 You could use counting on or counting back to complete these.

a $6 - 4 =$ ☐ b $8 - 3 =$ ☐ c $9 - 6 =$ ☐

d $10 - 7 =$ ☐ e $9 - 5 =$ ☐ f $10 - 8 =$ ☐

g $7 - 4 =$ ☐ h $10 - 4 =$ ☐ i $9 - 3 =$ ☐

2

a $5 - 3$ ☐ b $7 - 1$ ☐ c $10 - 3$ ☐

d $5 - 4$ ☐ e $6 - 3$ ☐ f $9 - 8$ ☐

g $9 - 4$ ☐ h $9 - 7$ ☐ i $10 - 6$ ☐ j $7 - 3$ ☐ k $8 - 5$ ☐

3

a 8 pegs – 4 pegs ☐

b 8 bugs – 2 bugs ☐

c 9 cats – 2 cats ☐

d 10 bears – 5 bears ☐

4

a

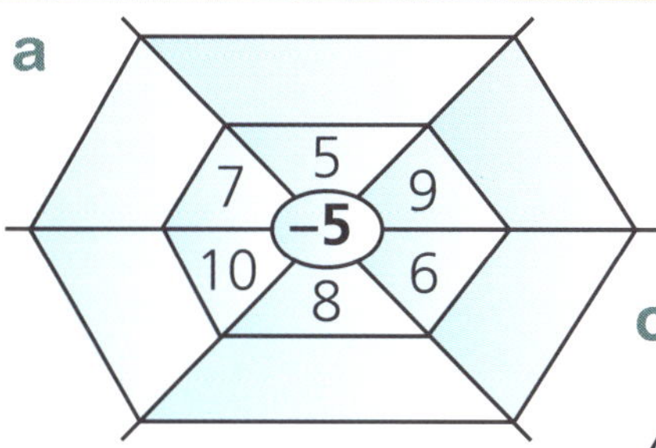

b

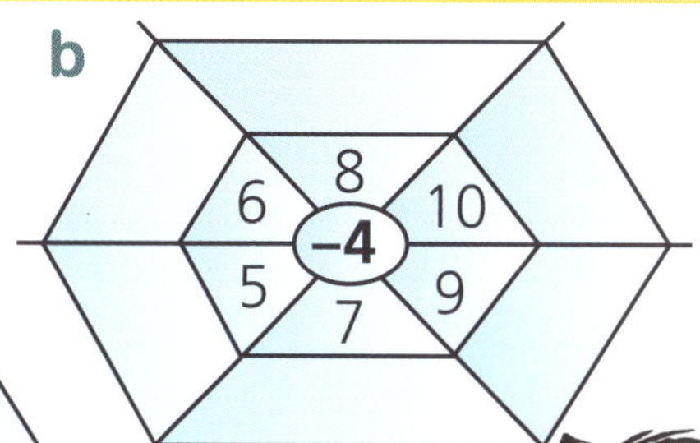

c

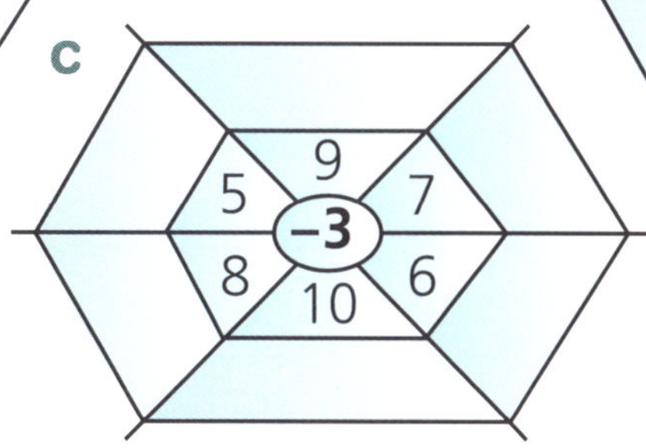

 • *AUSTRALIAN SIGNPOST MATHS 2* • ISBN 978 1 4886 2181 9

1C One Half

CONCEPT

1 Colour $\frac{1}{2}$ of each.

a

b

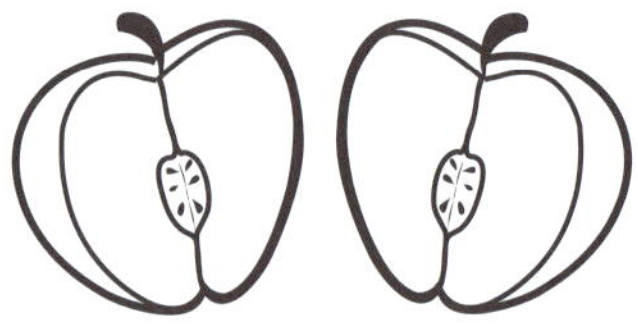

c

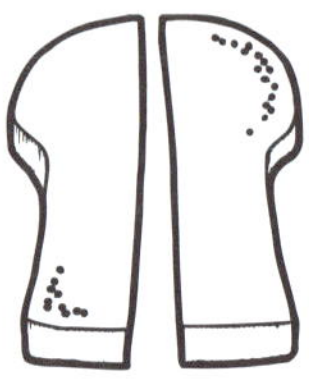

d

e

f

g

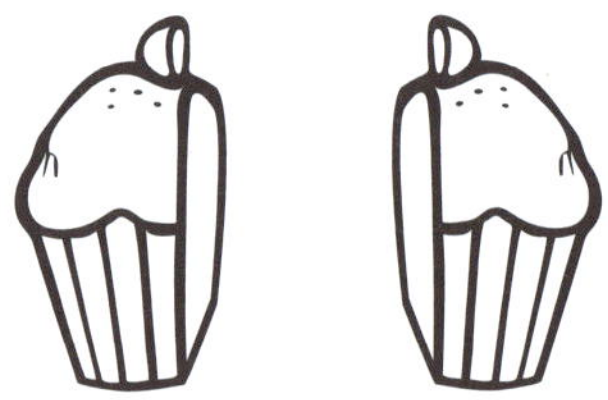

2 Draw a line to cut each object into halves. Colour one half of each.

a

b

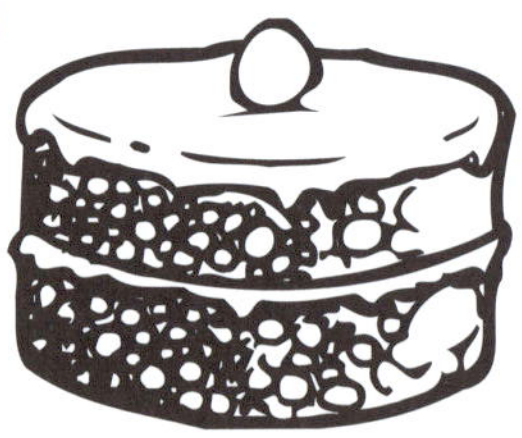

c

d

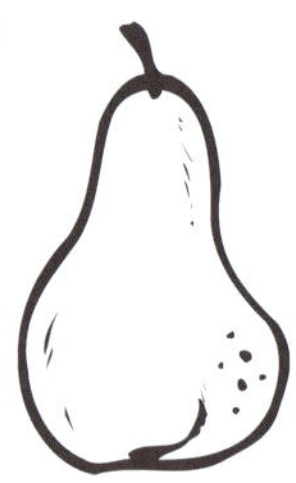

ACTIVITY

3 a Colour half of this group of balloons.

b Colour $\frac{1}{2}$ of each shape.

 • *AUSTRALIAN SIGNPOST MATHS 2* • ISBN 978 1 4886 2181 9

1D Parallel Lines

Parallel lines run in the same direction.
If continued, they do not meet.

❶ Colour the parallel lines red.

Line types

- broken lines
- sloping lines
- wavy lines
- zigzag lines
- crossed lines
- parallel lines

❷ What shapes can you see in this picture?

2A Ordinal Numbers

1 **a** Write the ordinal number for each dinosaur.

1st

Colour:

b the 3rd dinosaur green

c the 6th dinosaur red

d the 2nd dinosaur blue

e the 10th dinosaur purple

f the 4th dinosaur orange

g the 1st dinosaur pink

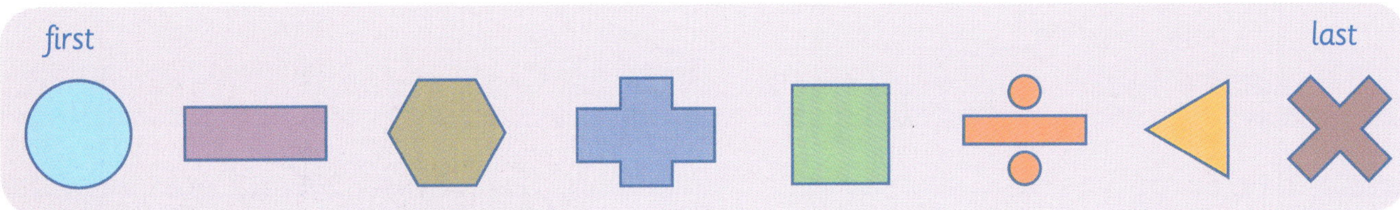

2 Write the ordinal number for:

a 

b 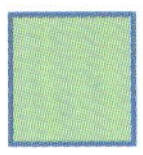

c 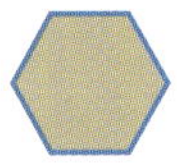

d 

3 Colour:

a 21st February red

b 2nd February green

c 19th February blue

d 26th February yellow

e 28th February purple

f 3rd February orange

February

Sun	Mon	Tues	Wed	Thurs	Fri	Sat
			1	2	3	4
5	6	7	8	9	10	11
12	13	14	15	16	17	18
19	20	21	22	23	24	25
26	27	28				

4 Write the date in February for:

a the last Saturday

b the first Monday

c the first Friday

d the last Sunday

 • *AUSTRALIAN SIGNPOST MATHS 2* • ISBN 978 1 4886 2181 9

2B Position Words

beside	next to	left	right	centre	middle
above	on	onto	on top of	below	beneath
under	underneath	bottom	back to back	in front of	upside down
behind	between	close to	forward	near	far
further away	up	high	low	in	inside

1 Use some of these words to give the position of:

a the emu

b the bird

c the rocks

d the kangaroo lying down

e the kangaroo standing up

f the long grass plant

2 Describe the position of animals in this picture.

The kangaroo (13) is in front of the tree.

2C Revision of Time

Both clocks show 2 o'clock.

Both clocks show half past 12.

1 Read the time and complete the sentence.
Write the time on the digital clock.

a 
I get up at half past ☐.
7:30

b 
Breakfast is at ☐ o'clock.
☐ : ☐

c 
Dinner is at half past ☐.
☐ : ☐

d 
I go to bed at ☐.
☐ : ☐

2 Match the analogue time to the correct digital time.

a
1:30 | 12:30

11:30 | 10:30

b
3:30 | 4:30

1:30 | 2:30

3 **a** Show these times on the digital clock.

School starts ☐ : ☐ Recess ☐ : ☐

Lunch ☐ : ☐ Home time ☐ : ☐

b Find out how many seconds there are in one minute. ☐

 ISBN 978 1 4886 2181 9

2D Ordering Capacities

1. Choose sets of three containers and measure the capacity of each. Record your results in the table below.

a

Container	Estimate	Count
	cups	cups
	cups	cups
	cups	cups

b

Container	Estimate	Count
	cups	cups
	cups	cups
	cups	cups

2. Order each set of containers above from holds least to holds most.

a 1 ______ 2 ______ 3 ______

b 1 ______ 2 ______ 3 ______

3. Tina recorded these results. Answer the following questions.

a Which container held 15 cups? ______

b Which container held 26 cups? ______

c Which container held less than the saucepan? ______

Container	Number	Comment
saucepan	15 cups	
bottle	11 cups	holds least
box	26 cups	holds most

d Which container held more than the saucepan? ______

e Order the containers from holds least to holds most.

1 ______ 2 ______ 3 ______

3A Counting Patterns

CONCEPT

Counting numbers to 30

1	2	3	4	5	6	7	8	9	10
11	12	13	14	15	16	17	18	19	20
21	22	23	24	25	26	27	28	29	30

Counting numbers from 100 to 130

101	102	103	104	105	106	107	108	109	110
111	112	113	114	115	116	117	118	119	120
121	122	123	124	125	126	127	128	129	130

The patterns for the numbers over 100 are like the patterns for numbers up to 100. Can you see a pattern?

7, 17, 27 ...
107, 117, 127 ...

Even numbers end in 2, 4, 6, 8 or 0.
Odd numbers end in 1, 3, 5, 7 or 9.

1 What number comes after:

a 110? ☐ b 114? ☐ c 118? ☐ d 121? ☐
e 123? ☐ f 125? ☐ g 127? ☐ h 129? ☐

2 What number comes before:

a 110? ☐ b 113? ☐ c 116? ☐ d 119? ☐
e 122? ☐ f 125? ☐ g 128? ☐ h 130? ☐

3 Complete the patterns. Check your answers with a calculator.

a 7, 17, 27, ☐, ☐, ☐, ☐, ☐ ...
b 48, 58, 68, ☐, ☐, ☐, ☐, ☐ ...
c 52, 62, 72, ☐, ☐, ☐, ☐, ☐ ...
d 39, 49, 59, ☐, ☐, ☐, ☐, ☐ ...
e 21, 31, 41, ☐, ☐, ☐, ☐, ☐ ...

 ISBN 978 1 4886 2181 9

3B Counting

CONCEPT

1	2	3	4	5	6	7	8	9	10
11	12	13	14	15	16	17	18	19	20
21	22	23	24	25	26	27	28	29	30
31	32	33	34	35	36	37	38	39	40
41	42	43	44	45	46	47	48	49	50
51	52	53	54	55	56	57	58	59	60
61	62	63	64	65	66	67	68	69	70
71	72	73	74	75	76	77	78	79	80
81	82	83	84	85	86	87	88	89	90
91	92	93	94	95	96	97	98	99	100

101	102	103	104	105	106	107	108	109	110
111	112	113	114	115	116	117	118	119	120
121	122	123	124	125	126	127	128	129	130

1. Count aloud.
   - a Start at 82 and count by 2s to 126.
   - b Start at 73 and count by 2s to 109.
   - c Start at 25 and count by 5s to 130.
   - d Start at 43 and count by 10s to 123.

2. Complete the number patterns and describe each.
   - a 95, 100, 105, ___, ___, ___, ___, ___
   - b 60, 70, 80, ___, ___, ___, ___, ___
   - c 130, 128, 126, ___, ___, ___, ___, ___

ICT

3. Use a calculator to complete these number patterns. Discuss.
   - a 789 [+] 2 [=] [=] [=] [=] [=]

     Display shows: 791, ___, ___, ___, ___
   - b 456 [+] 5 [=] [=] [=] [=] [=]

     Display shows: 461, ___, ___, ___, ___

3C One Quarter

CONCEPT

1 Colour $\frac{1}{4}$ of each.

a 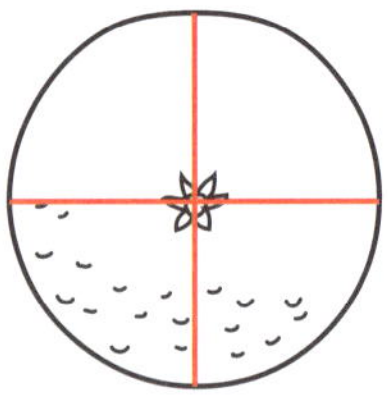

b 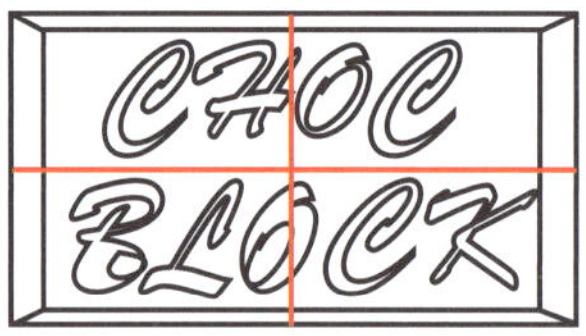

c 

d 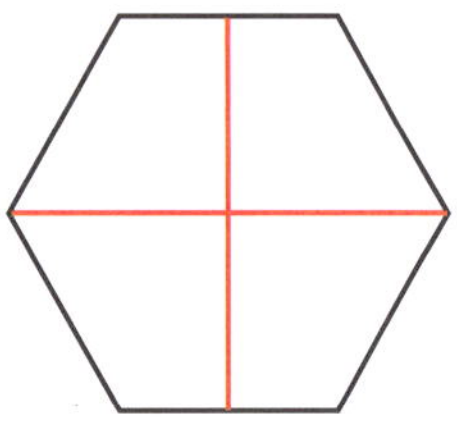

e 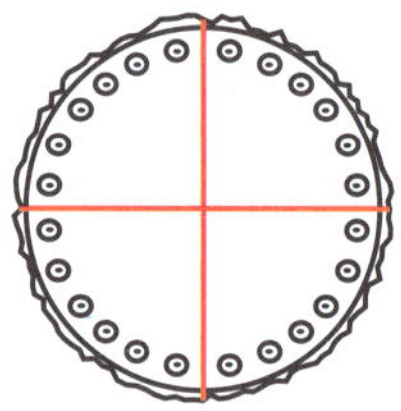

f 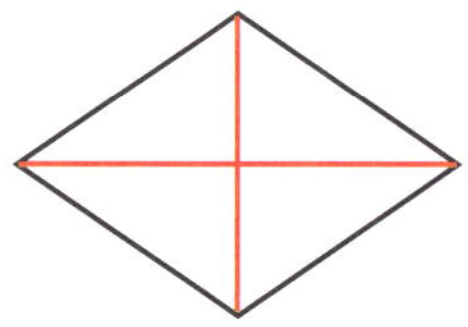

2 Draw two lines to cut each shape into quarters.

a 

b 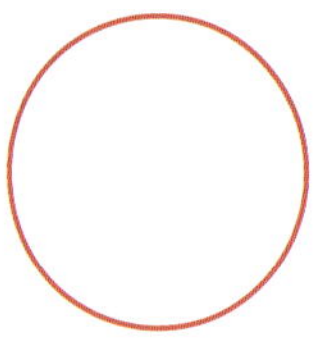

c 

d 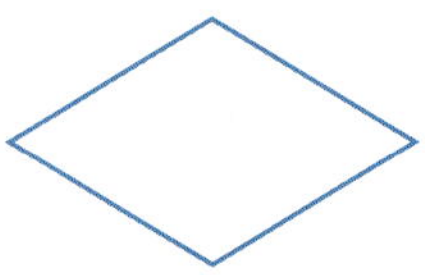

e 

INVESTIGATION

3 I have 4 ten-cent coins.

a What is the total value? ______

b What is $\frac{1}{2}$ of this? ______

c What is $\frac{1}{4}$ of this? ______

4 Find $\frac{1}{2}$ and $\frac{1}{4}$ of other amounts of coins that have the same value.

 • *AUSTRALIAN SIGNPOST MATHS 2* • ISBN 978 1 4886 2181 9

Describing 3D Objects

1 Complete:

		Slides	Rolls	Stacks	Sinks	Floats
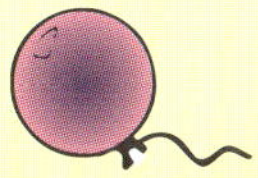	Balloon	—	✓	—	—	✓
	Nut					
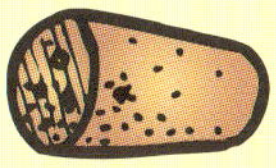	Cork					
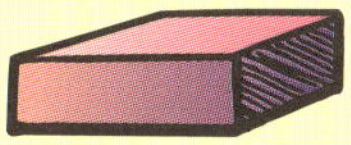	Block					
	Can					
	Marble					

2 Colour the objects that match the description.

6 square faces	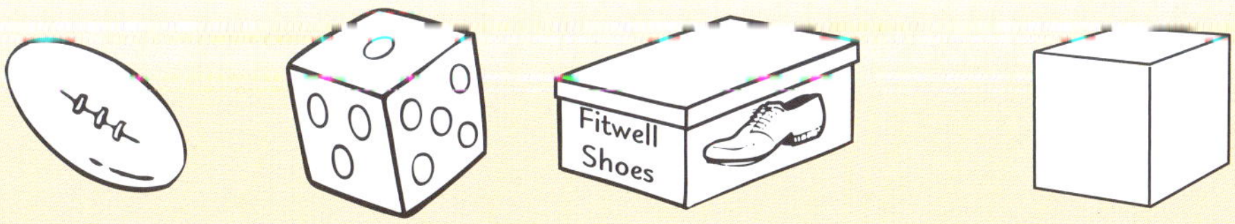
No corners	

INVESTIGATION

Find objects in the classroom to complete the table.

Object	Curved Surface(s)	Flat Surface(s)
	✓	
		✓
	✓	✓

4A Addition to 20

1 Use counters or counting on to complete each number sentence.

a 4 + 9 = ☐ b 6 + 8 = ☐ c 7 + 9 = ☐

d 5 + 7 = ☐ e 9 + 8 = ☐ f 7 + 7 = ☐

g 10 + 8 = ☐ h 9 + 10 = ☐ i 8 + 8 = ☐

2

a 5 + 3 = ☐ b 8 + 4 = ☐ c 7 + 8 = ☐

d 5 + 9 = ☐ e 9 + 9 = ☐ f 7 + 6 = ☐

g 6 + 7 = ☐ h 10 + 10 = ☐ i 12 + 3 = ☐

We read this as 6 + 4 = 10.

6
+ 4
10

How many possible number sentences for 12 are there?

8 + 7 = 7 + 8

3 Write a different true number sentence each time.

a 12 = ☐ + ☐
12 = ☐ + ☐
12 = ☐ + ☐
12 = ☐ + ☐

b 15 = ☐ + ☐
15 = ☐ + ☐
15 = ☐ + ☐
15 = ☐ + ☐

INVESTIGATION

Doubles

Roll two dice. Add the numbers.

Write a number sentence to show this.

☐ + ☐ = ☐

© PEARSON AUSTRALIA 2018 • *AUSTRALIAN SIGNPOST MATHS 2* • ISBN 978 1 4886 2181 9

4B Addition to 20

8 + 5 Start at 8 and count on.

7 8 9 10 11 12 13

$$\begin{array}{r} 8 \\ +\ 5 \\ \hline 13 \end{array}$$

We read this as 8 + 5 = 13.

1 Use the number line to count on from the largest number.

5 6 7 8 9 10 11 12 13 14 15

Learn your addition tables.

a 7 + 6 = ☐	**b** 4 + 7 = ☐	**c** 3 + 9 = ☐	**d** 8 + 4 = ☐	**e** 5 + 6 = ☐
f 4 + 8 = ☐	**g** 9 + 2 = ☐	**h** 6 + 8 = ☐	**i** 2 + 9 = ☐	**j** 7 + 8 = ☐

2 Use counters to find numbers for the windows.

17	19	18	16	15	20
☐ + ☐	☐ + ☐	☐ + ☐	☐ + ☐	☐ + ☐	☐ + ☐

3 **a** An ant has 6 legs.
Find different ways its legs can be arranged.

☐ + ☐ + ☐ = 6

☐ + ☐ + ☐ = 6

b An octopus has 8 legs.
Find different ways its legs can be arranged.

☐ + ☐ + ☐ = 8

☐ + ☐ + ☐ = 8

 • *AUSTRALIAN SIGNPOST MATHS 2* • ISBN 978 1 4886 2181 9

4C Comparing Masses

1 a Handle two objects like these and estimate which one is heavier.

I estimate the ______ to be

heavier than the ______.

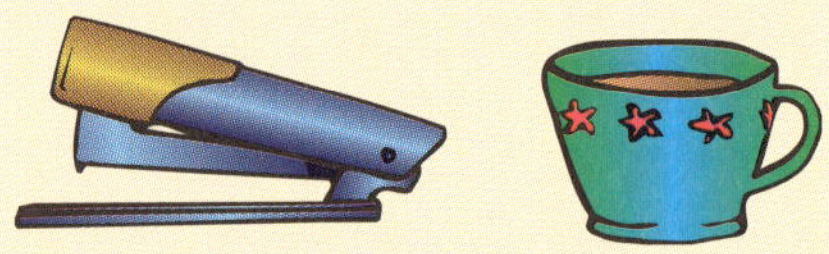

b Use blocks and a balance scale to find the mass of each object.

(You could use marbles as a unit.)

The ______ has a mass equal to ______ blocks.

The ______ has a mass equal to ______ blocks.

2 Choose three pairs of objects in the classroom.

- Estimate which object is heavier in each pair by hefting.
- Check by using a balance scale and marbles.
- Complete this chart.

Estimate by Hefting	Number of Units	Was Your Estimate Correct?
______ is heavier than ______	______ ______	Yes No
______ is heavier than ______	______ ______	Yes No
______ is heavier than ______	______ ______	Yes No

Lightest to Heaviest

Choose eight objects. Order them according to mass.

4D Balance Scales

Using scales is like hefting.

The scales are balanced.

1 One book balances ☐ apples.

2 Which objects are lighter or heavier than a place-value hundred block?

basketball

stapler
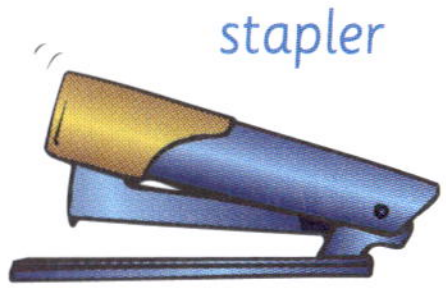

book

scissors

tennis ball

brick
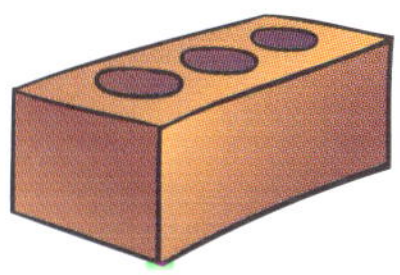

lunch box

Lighter	Heavier

3 Balance and record.

Object		Unit of Measure		
	cup	☐	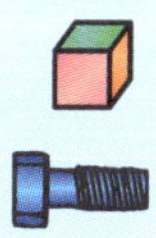	blocks or bolts
	scissors	☐		pencils
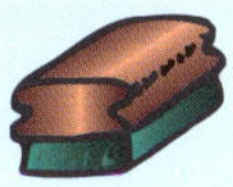	duster	☐		blocks

4 Find 3 objects that have nearly the same mass as a cup.

5A Modelling Numbers

CONCEPT

The number 115 can be represented using many different materials.

calculator

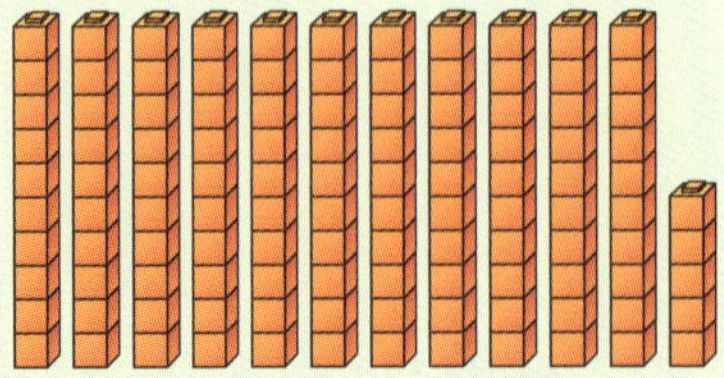

connecting cubes

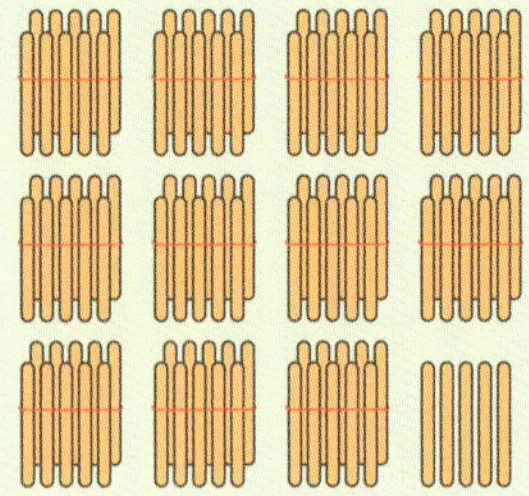

craft sticks

1 Write the number modelled.

a

b

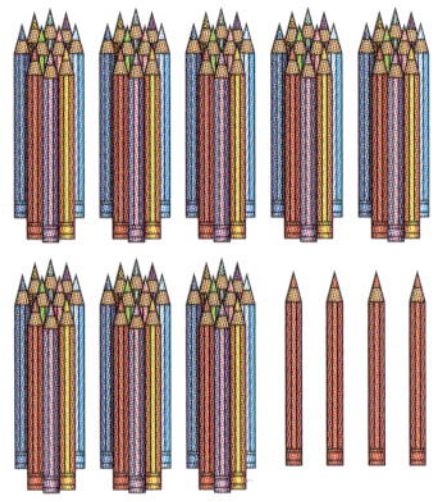

c

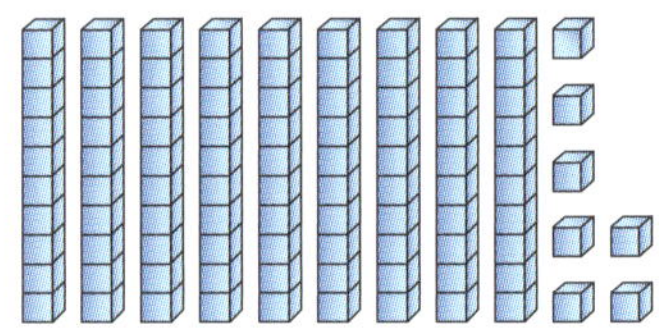

d

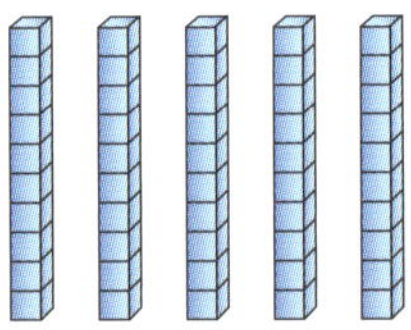

e

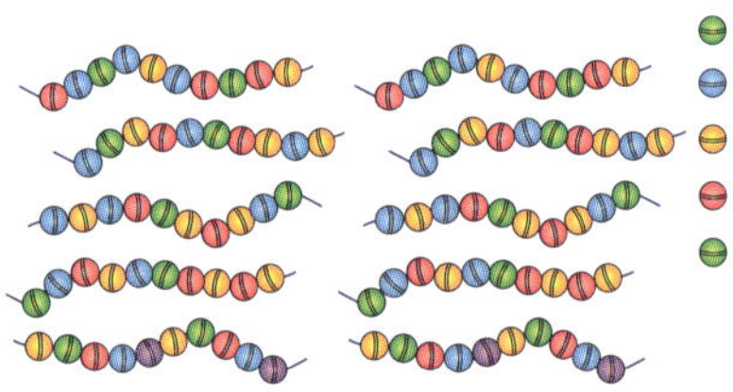

f

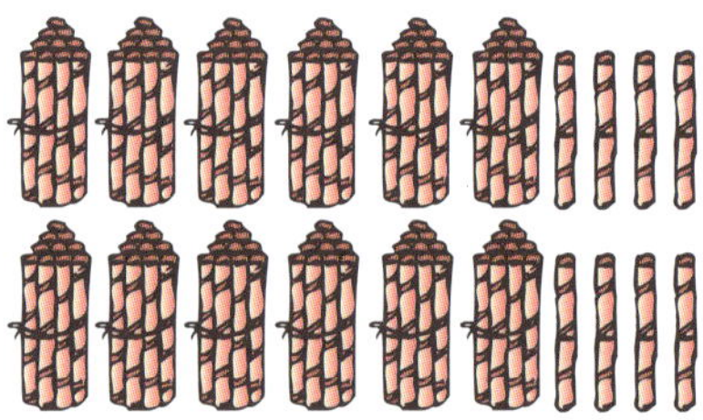

2 Count aloud from 89 to 130 from memory.

3 Count aloud from 75 to 135 by 5s, and then by 10s.

FUN SPOT

- Every 10th bead around this page is coloured red.
- Using a green pencil, colour bead number:
  - 100
  - 115
  - 127
  - 49
  - 149
- Use a calculator to make the numbers above.

START

 • *AUSTRALIAN SIGNPOST MATHS 2* • ISBN 978 1 4886 2181 9

5B Doubling and Near Doubling

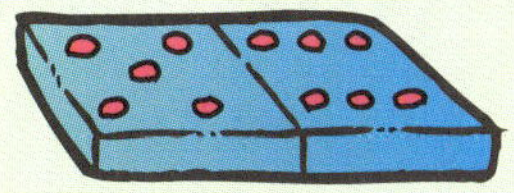

CONCEPT

When you **double** a number, the answer will be **even**.

1 Complete each number sentence.

a double 3

☐ + ☐ = ☐

b double 6

☐ + ☐ = ☐

c double 1

☐ + ☐ = ☐

d double 7

☐ + ☐ = ☐

e double 9

☐ + ☐ = ☐

f double 8

☐ + ☐ = ☐

2

a 2 + 2 = ☐

b 4 + 4 = ☐

c 0 + 0 = ☐

d 5 + 5 = ☐

e 10 + 10 = ☐

3

5C 2D Shapes

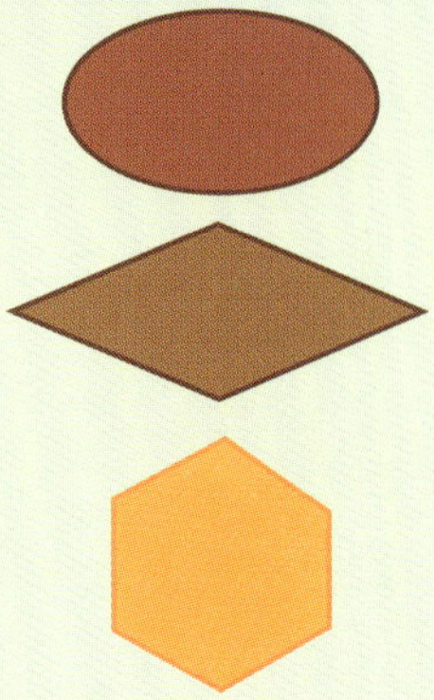

Oval: one curved side

Rhombus: all sides equal

Hexagon: six straight sides

1. In this picture, find how many:
   - a circles
   - b triangles
   - c squares
   - d hexagons
   - e ovals
   - f rhombuses

   Discuss their features.

2. a Draw a line to match each shape with the hole it fits.
   b Write the name of each shape in the box below.

5D Informal Units of Length

1 Use the object named as a measuring unit to find the length of your desk.

Unit Used	Length of Your Desk
This book	☐ books
Your pen	☐ pens
Finger lengths	☐ fingers

2 Use your hand span to measure the length of each object.

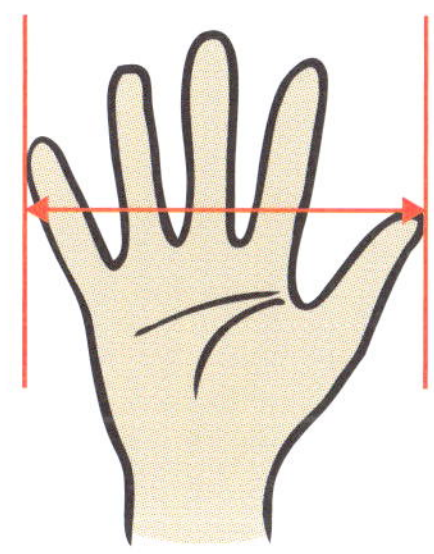

Your desk	☐ hand spans
Teacher's desk	☐ hand spans

Which is longer? ☐

Window	☐ hand spans
Door	☐ hand spans

Which is shorter? ☐

3 Estimate, then use craft sticks to measure these lengths.

a

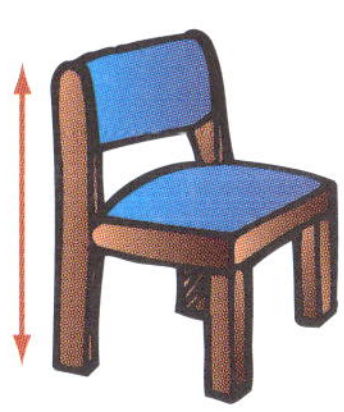

Estimate ☐

Measure ☐

b

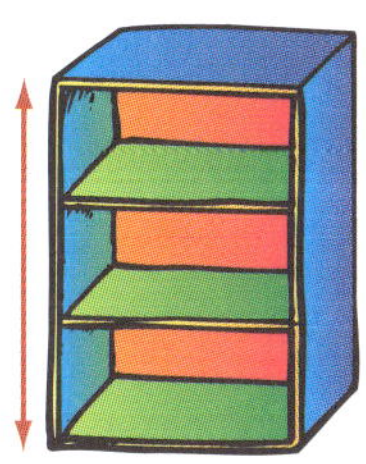

Estimate ☐

Measure ☐

c

Estimate ☐

Measure ☐

Use string to compare the length of objects in the room.

☐ is longer than ☐.

☐ is longer than ☐.

 • *AUSTRALIAN SIGNPOST MATHS 2* • ISBN 978 1 4886 2181 9

6A Groups and Rows

CONCEPT

You can count equal groups to find the total.

1

4 groups of 2 makes ☐.

2

3 rows of 5 makes ☐.

3

6 groups of 2 makes ☐.

ACTIVITY

4 Draw 3 rows of 4 balls.

5 Draw 5 groups of 3 stars.

Use counters to model each problem on this page.

 • *AUSTRALIAN SIGNPOST MATHS 2* • ISBN 978 1 4886 2181 9

6B Groups and Rows

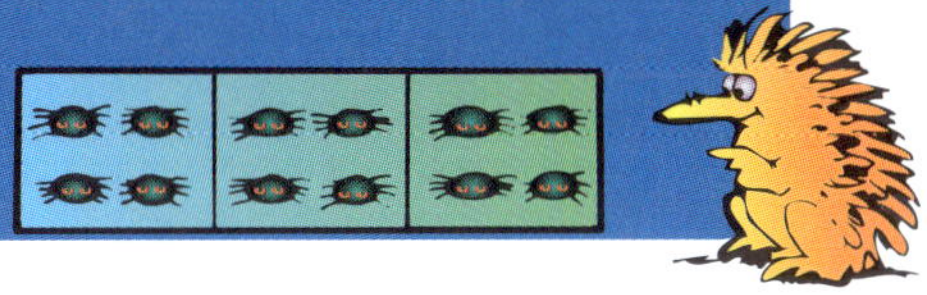

CONCEPT

4 groups of 4 = 16

1 Complete these.

a

☐ + ☐ + ☐

3 groups of 3 = ☐

b

☐ + ☐ + ☐

3 groups of 4 = ☐

c

6 groups of 3 = ☐

d

5 rows of 4 = ☐

e

☐ rows of ☐ = ☐

ACTIVITY

2 Use counters to solve these problems.

a 3 groups of 5 balls = ☐ balls

b 2 rows of 7 trees = ☐ trees

c 3 groups of 8 ants = ☐ ants

d 4 rows of 6 people = ☐ people

6C Analogue Time

CONCEPT

At quarter past, the small hand is pointing just past the number and the big hand is pointing to the 3.

15 past 7
a quarter past 7

At quarter to, the small hand is almost pointing to the number and the big hand is pointing to the 9.

45 past 7
a quarter to 8

1. Write the time shown on each face using **quarter to** or **quarter past**.

a 

b 

c 

d 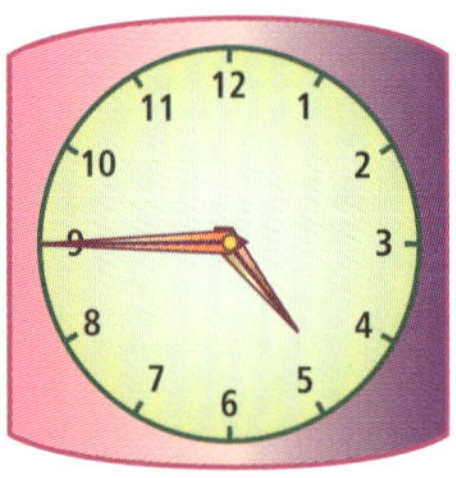

e 

f 

2. On each face show the time given.

a 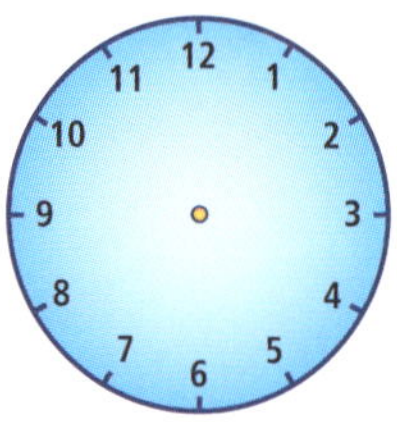
a quarter past 2

b 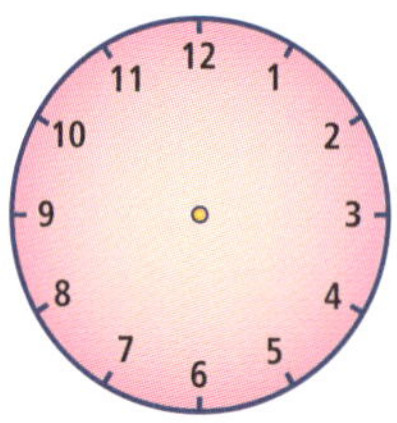
a quarter to 4

c 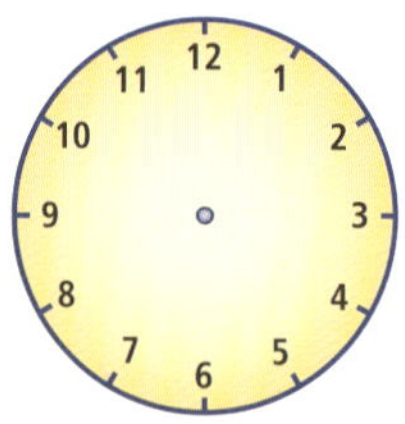
a quarter to 10

3. At a quarter ______ the hour, the small hand is pointing just past the hour number and the big hand is pointing to the ______.

6D Analogue Time

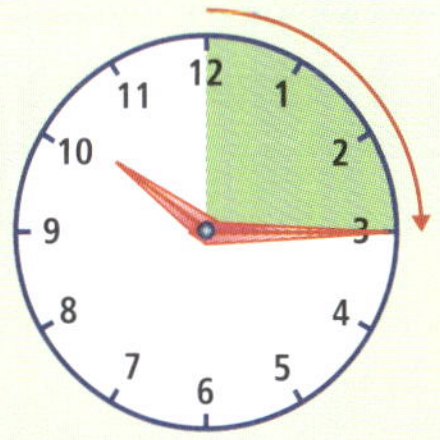

The big hand has moved a quarter of the way around the clock.

a quarter past 10

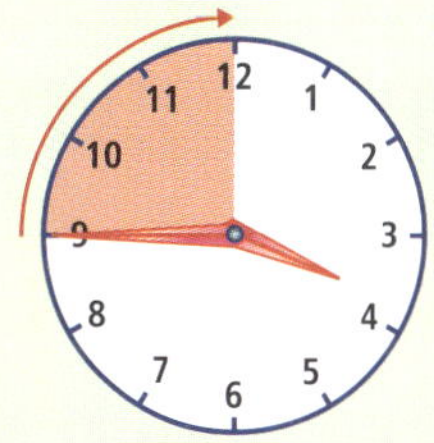

The big hand has a quarter of the clock to go before it is 4 o'clock.

a quarter to 4

1 Write the time shown.

a

b

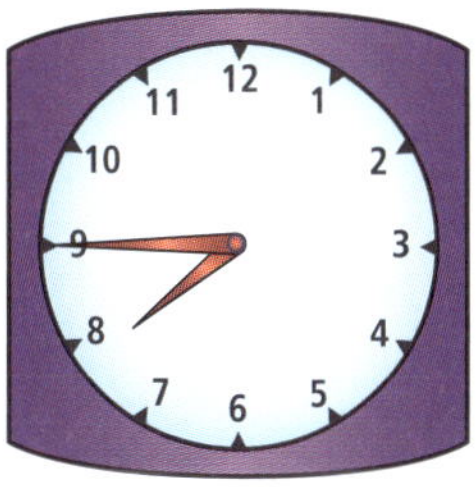

c

d

e

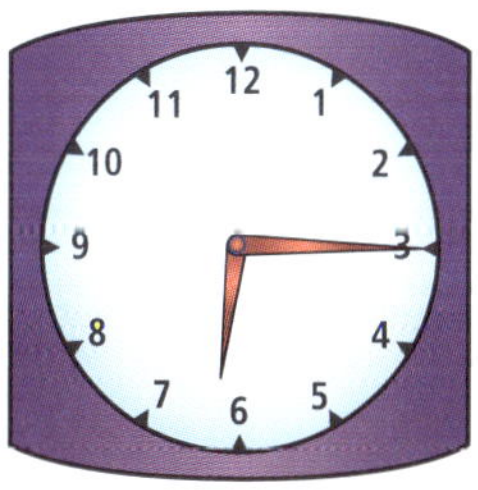

f

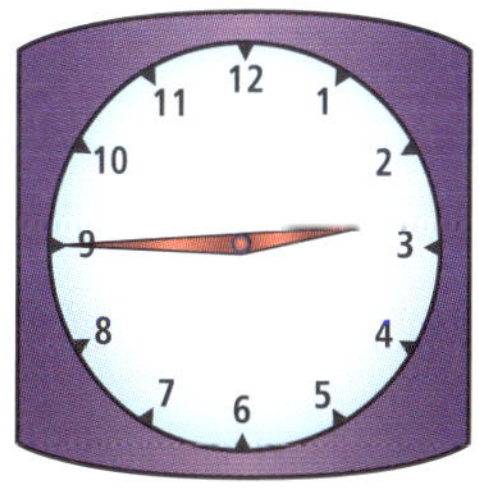

2 Show the time.

a

a quarter to 1

b

a quarter past 7

c

a quarter to 5

3 Find how many minutes are in:

a a quarter of an hour

b three quarters of an hour

7A Sharing

When we share, each person gets the same number of items.

1. Share 9 counters among these boxes.

   How many are there in each box?

2. Share 8 counters among these boxes.

   How many are there in each box?

3. Is this a fair share?

Circle the group with the unequal share.

4. Draw extra flowers to make these groups equal.

5. Draw extra fish to make these groups equal.

Sharing

When sharing fairly, each person is given the same amount.

❶ Share 6 counters among 3 boxes.

One share = ☐

❷ Share 8 blocks between 2 groups.

One share = ☐

❸ Share 10 blocks between 2 groups.

One share = ☐

❹ Share 12 counters:

a between 2 people

One share = ☐

b among 3 people

One share = ☐

c among 4 people

One share = ☐

d among 6 people

One share = ☐

7C Graphing the Weather

1 Mrs Johnson's class graphed the weather for May.

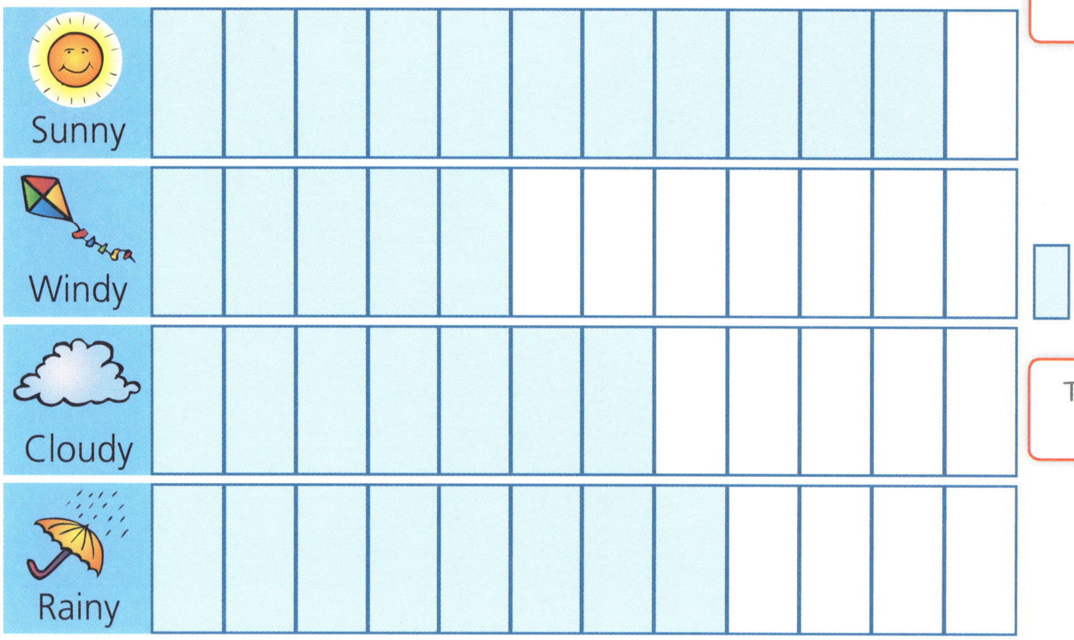

a How many days were:

sunny? ☐ windy? ☐ cloudy? ☐ rainy? ☐

b Were there more sunny days than rainy days? ☐

c How many more days were sunny than cloudy? ☐

d What kind of weather was most common in May? ☐

e What kind of weather was least common in May? ☐

f How many days are there in the month of May? ☐

2 Record the weather for one week and write about it.

Sun	Mon	Tues	Wed	Thurs	Fri	Sat

 • *AUSTRALIAN SIGNPOST MATHS 2* • ISBN 978 1 4886 2181 9

7D Thinking about Graphs

1 A

B

Which of these do you think is the better picture graph?

Why do you think so?

2 During the year these children were given stickers:

a How many stickers was Alana given?

b Who was given the least?

c How many stickers were given altogether?

d Complete the table.

Name	Number
Alana	
Deepak	
Brendon	

3 Use Question 1 to finish this graph.

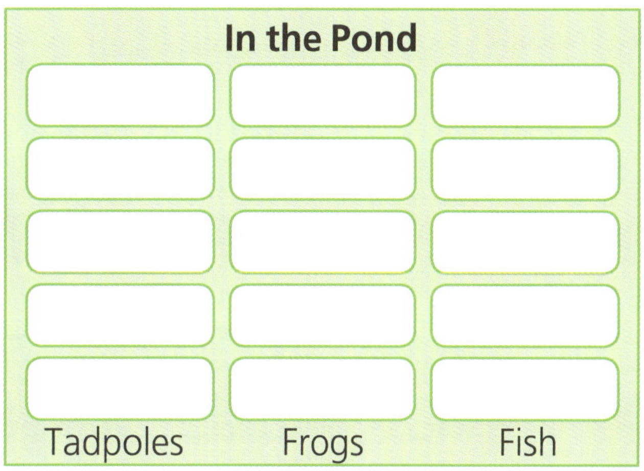

8A Subtraction to 20

Billy had 12 apples.

He ate 5.

How many did he have left?

12 − 5 = 7

$$\begin{array}{r} 12 \\ -\ 5 \\ \hline 7 \end{array}$$

CONCEPT

1 **a** 10 bottles
8 broken
How many left?
☐ − ☐ = ☐

b 12 flowers
7 picked
How many left?
☐ − ☐ = ☐

c 9 cakes
6 eaten
How many left?
☐ − ☐ = ☐

d 15 balls
6 lost
How many left?
☐
− ☐
☐

e 13 pencils
8 broken
How many left?
☐
− ☐
☐

f 14 mice
7 caught
How many left?
☐
− ☐
☐

g 16 nuts
9 eaten
How many left?
☐
− ☐
☐

FUN SPOT

2 Challenge a friend to answer these questions quickly.

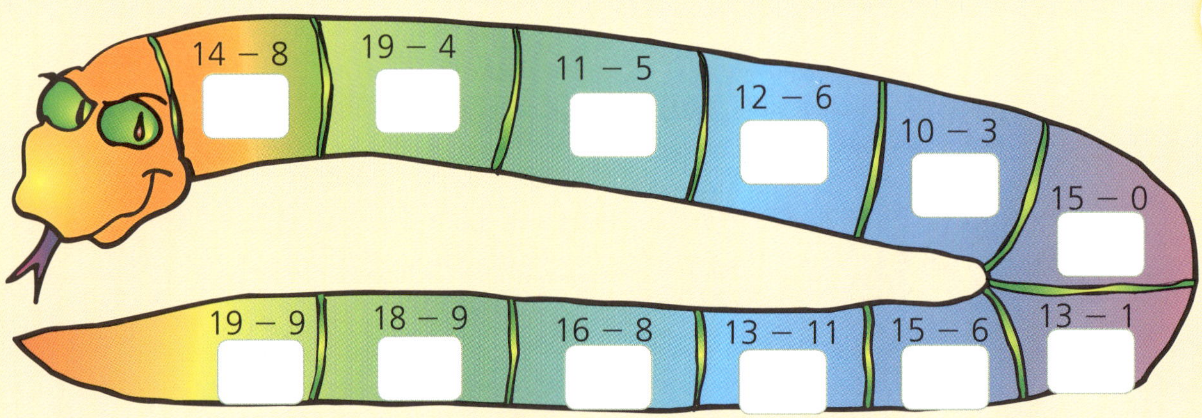

  • *AUSTRALIAN SIGNPOST MATHS 2* • ISBN 978 1 4886 2181 9

8B Addition to 20

1 Draw a picture and complete the number sentences.

a Milly got 8 pencils from Dad, 6 from Mum and 5 from Bess. How many did she get altogether?

☐ + ☐ + ☐ = ☐

b Ben had 6 DVDs. Dom gave him 4 DVDs and Lee gave him 8. How many DVDs does Ben have altogether?

☐ + ☐ + ☐ = ☐

2 Count on or use doubles to complete.

a 3 + 4 + 3 = ☐	**b** 10 + 2 + 2 = ☐	**c** 6 + 1 + 6 = ☐
d 4 + 3 + 5 = ☐	**e** 8 + 4 + 2 = ☐	**f** 6 + 4 + 4 = ☐
g 4 + 3 + 4 = ☐	**h** 1 + 6 + 4 = ☐	**i** 3 + 6 + 2 = ☐

3 Write three number sentences for each total.

a
☐ + ☐ + ☐ = 15
☐ + ☐ + ☐ = 15
☐ + ☐ + ☐ = 15

b
☐ + ☐ + ☐ = 19
☐ + ☐ + ☐ = 19
☐ + ☐ + ☐ = 19

c
☐ + ☐ + ☐ = 16
☐ + ☐ + ☐ = 16
☐ + ☐ + ☐ = 16

d
☐ + ☐ + ☐ = 18
☐ + ☐ + ☐ = 18
☐ + ☐ + ☐ = 18

How many number sentences can you find that add up to 20?

Discuss which number above would have the most possible number sentences.

© PEARSON AUSTRALIA 2018 • *AUSTRALIAN SIGNPOST MATHS 2* • ISBN 978 1 4886 2181 9

8C Half of a Group

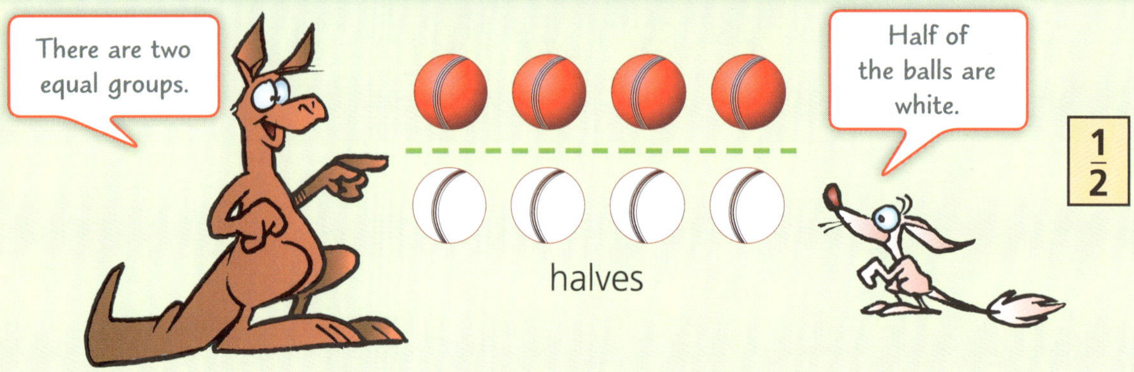

CONCEPT

1. Circle the groups that are divided into halves. Colour one half of each group that you circled.

a 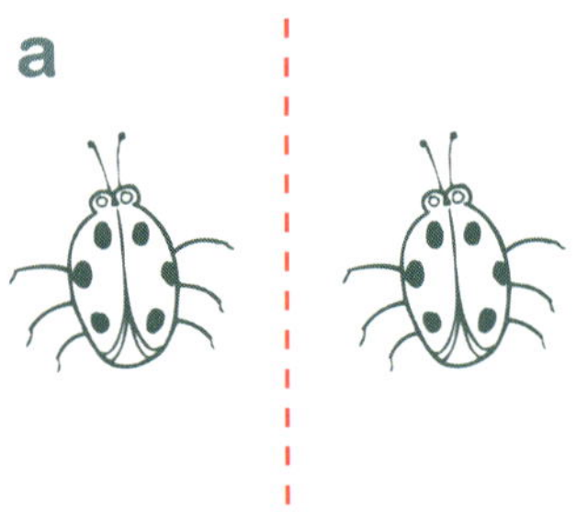

b 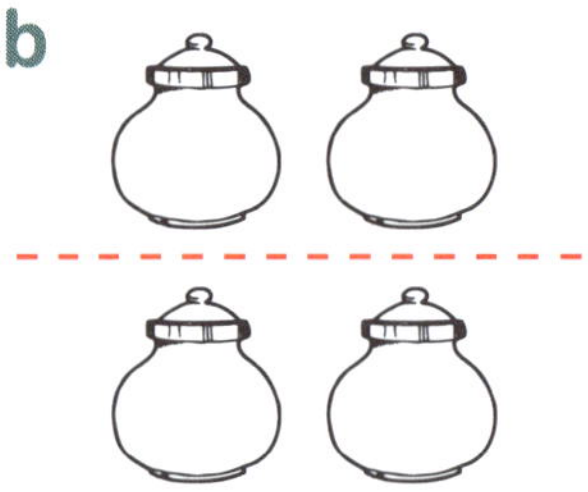

c 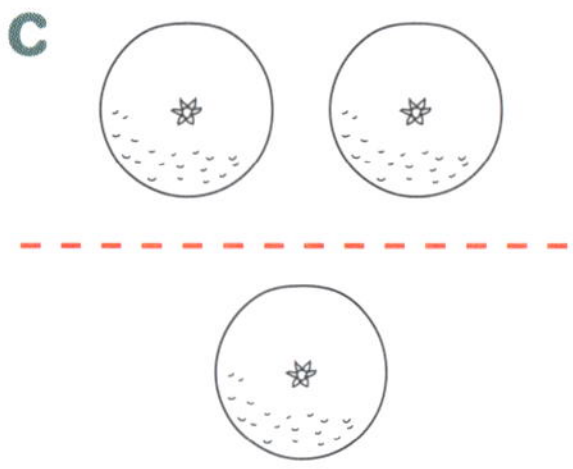

d 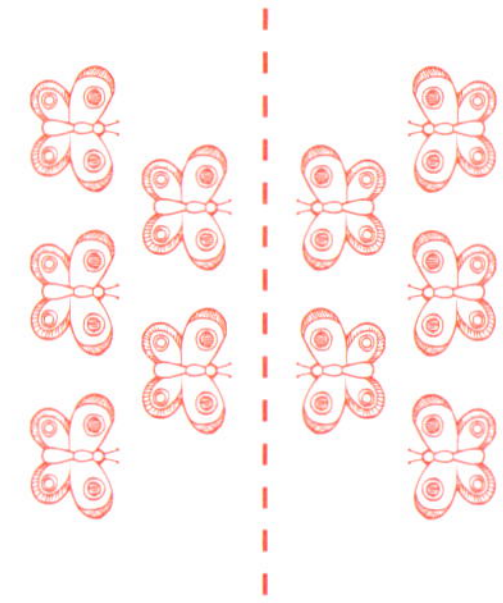

e 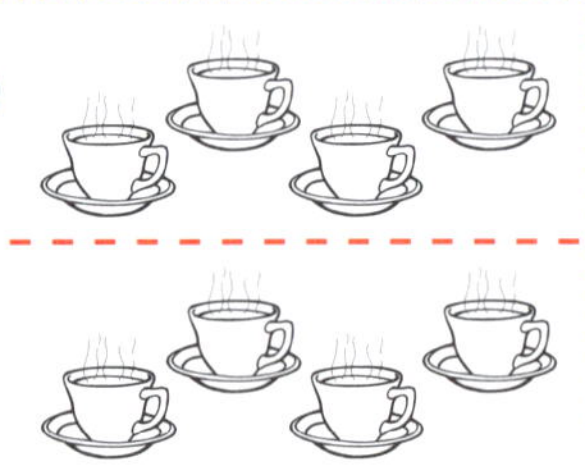

f 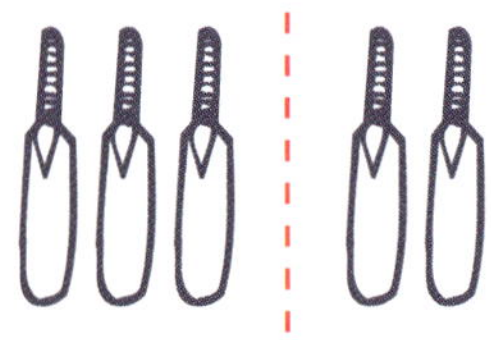

g 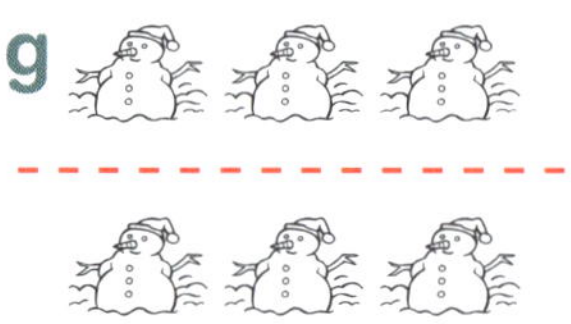

h 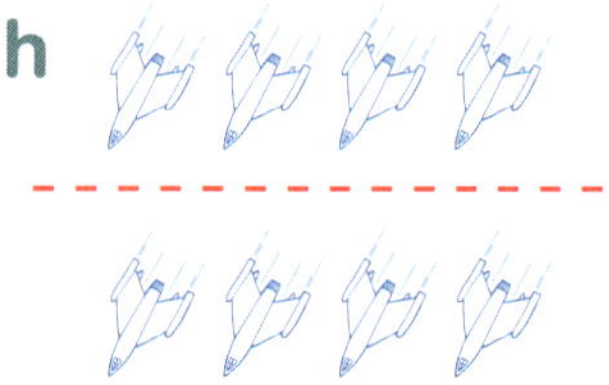

2. Circle half of each group.

a 

b 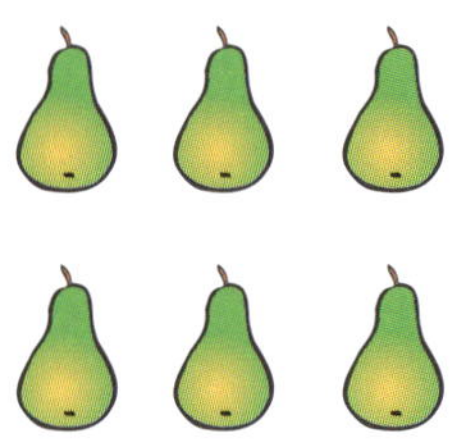

c 

d 

e 

f 

g 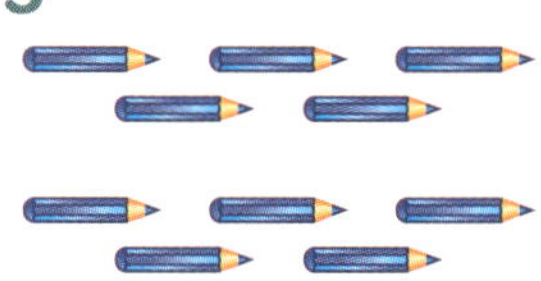

h 

© PEARSON AUSTRALIA 2018 • *AUSTRALIAN SIGNPOST MATHS 2* • ISBN 978 1 4886 2181 9

8D Quarter of a Group

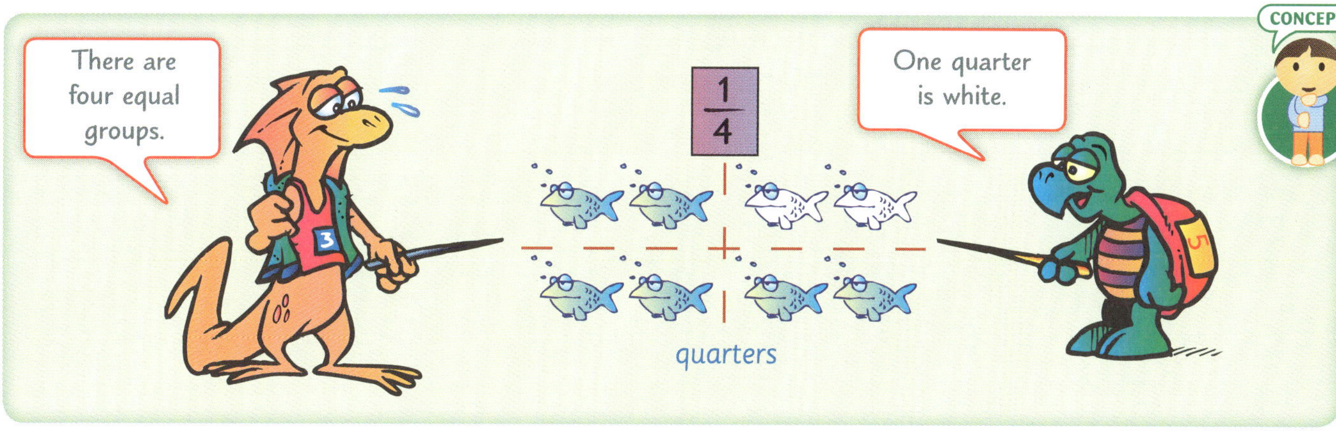

1 Circle the groups that are divided into quarters and colour one quarter.

a
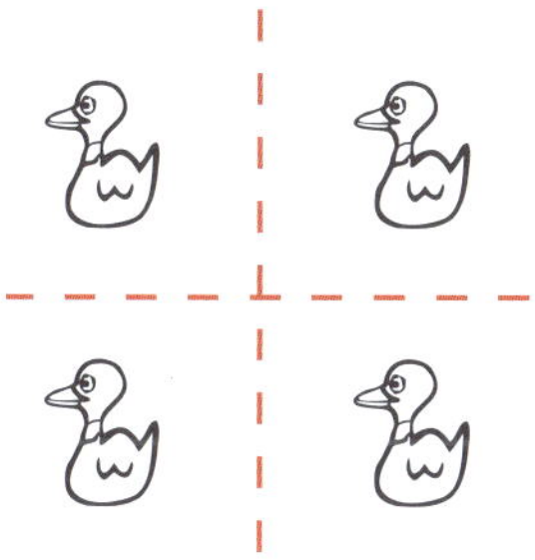

b
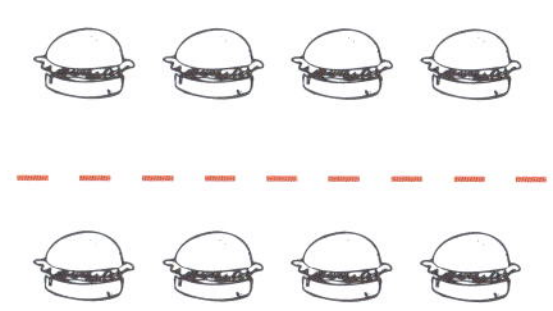

c
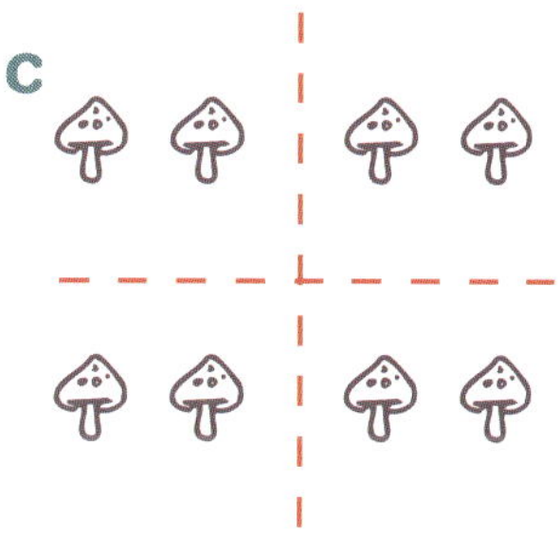

d

e
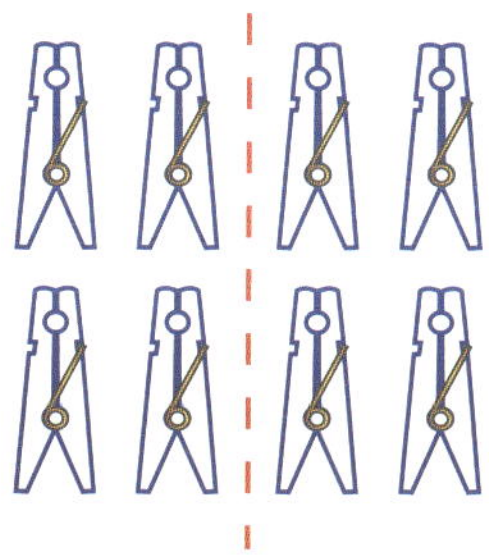

f
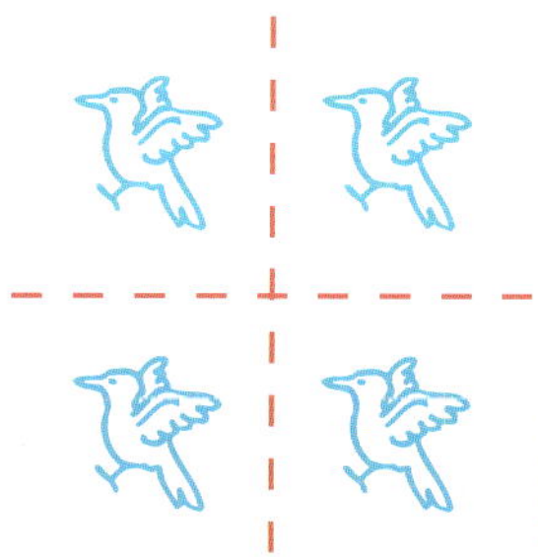

g

h
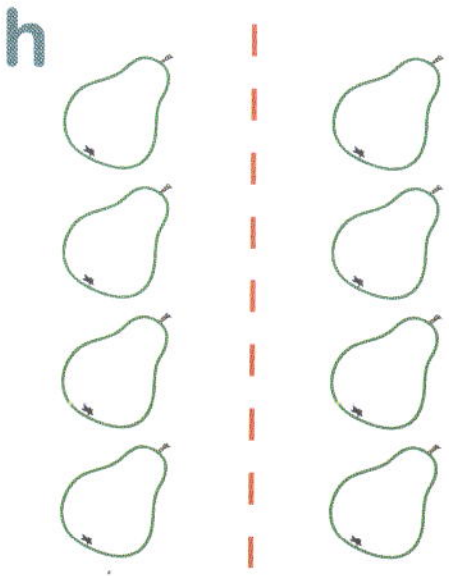

2 Circle one quarter of each group.

a

b

c

d

e
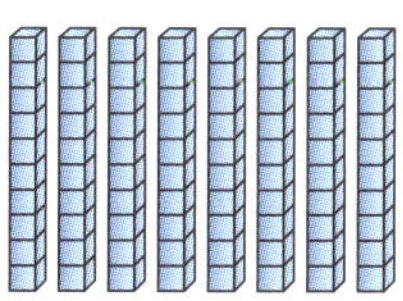

f

g
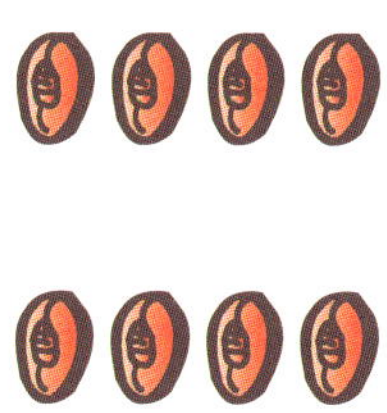

h

9A Numbers to 150

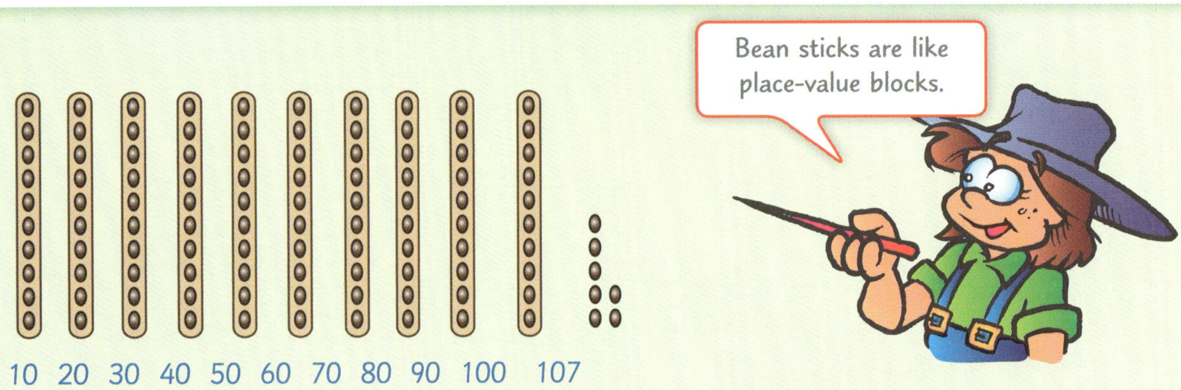

1 Write these numerals.

a

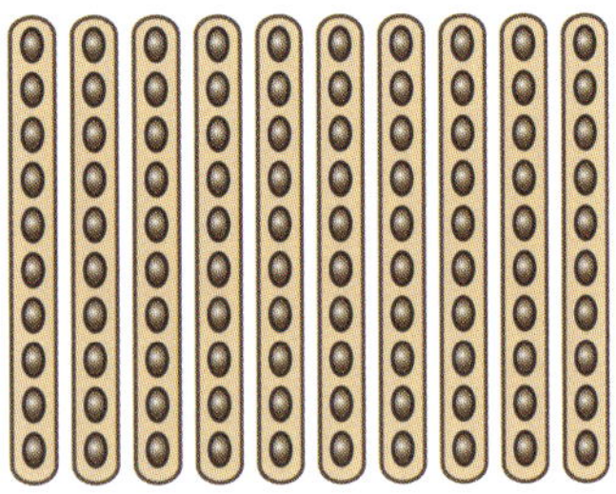

Hundreds | Tens | Ones

b

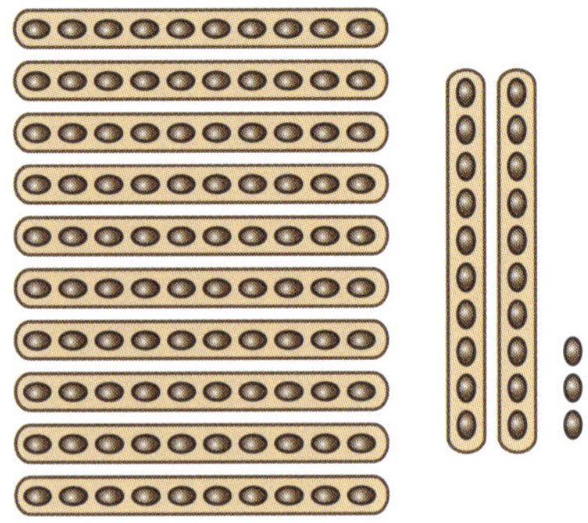

Hundreds | Tens | Ones

c

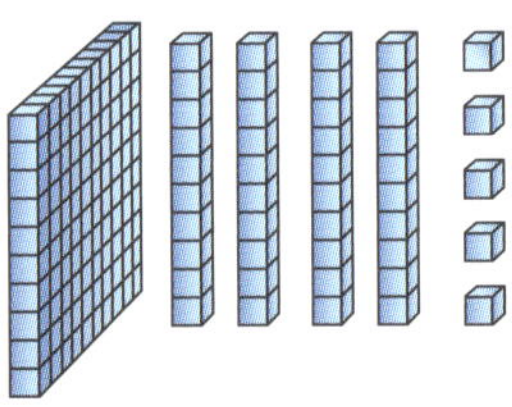

Hundreds | Tens | Ones

d

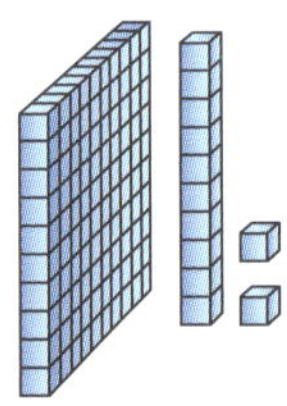

Hundreds | Tens | Ones

e

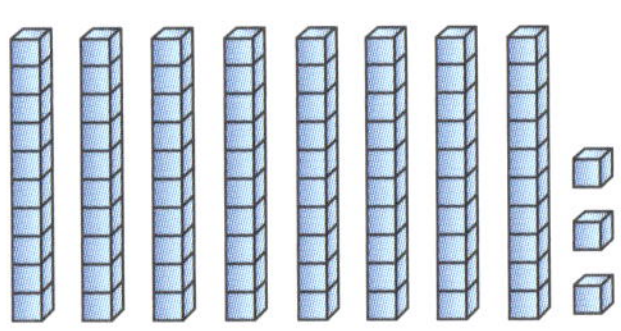

Hundreds | Tens | Ones

f

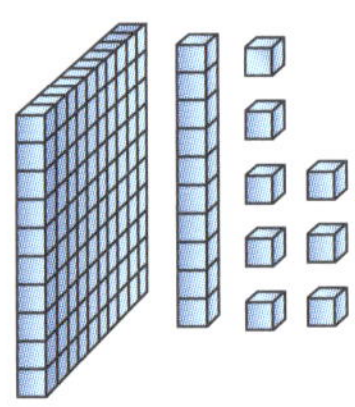

Hundreds | Tens | Ones

- Use bean sticks or place-value blocks to make these numbers.
  - 106
  - 98
  - 117
  - 123
  - 66
  - 100
- Explain your answers to a friend.

 • *AUSTRALIAN SIGNPOST MATHS 2* • ISBN 978 1 4886 2181 9

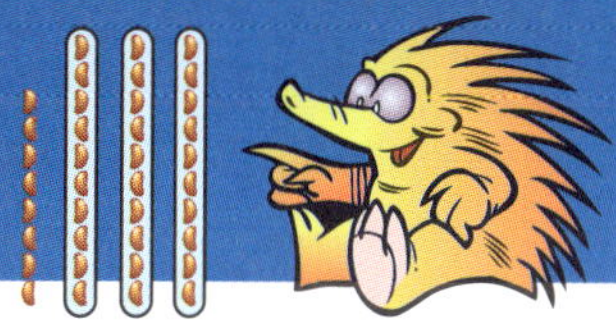

1 Complete the numeral expander and write the numeral.

a

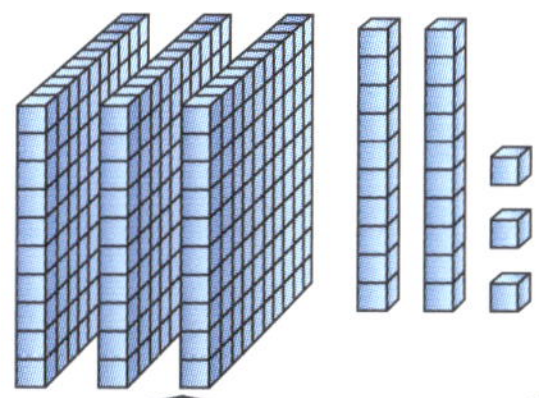

= ☐

☐ Hundreds ☐ Tens ☐ Ones

b

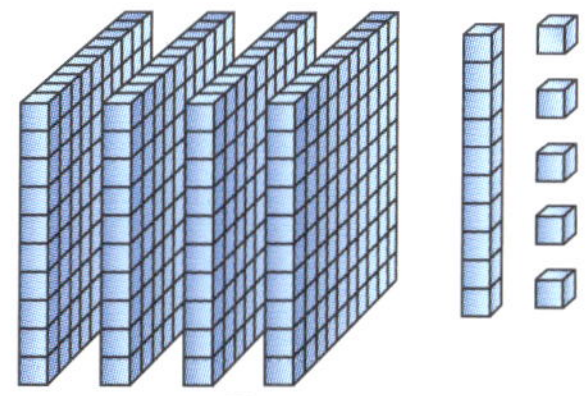

= ☐

☐ Hundreds ☐ Tens ☐ Ones

c

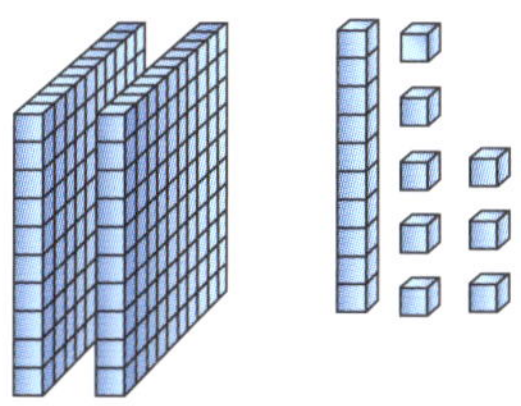

= ☐

☐ Hundreds ☐ Tens ☐ Ones

d

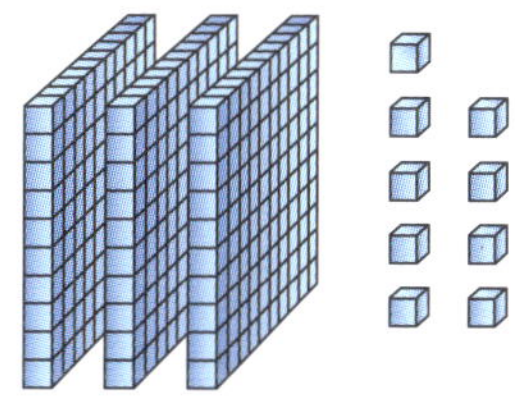

= ☐

☐ Hundreds ☐ Tens ☐ Ones

2 Write the next number.

a 213 ☐	**b** 407 ☐	**c** 728 ☐	**d** 699 ☐
e 817 ☐	**f** 202 ☐	**g** 909 ☐	**h** 325 ☐

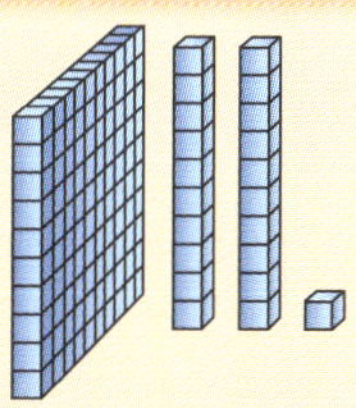

Place-value blocks

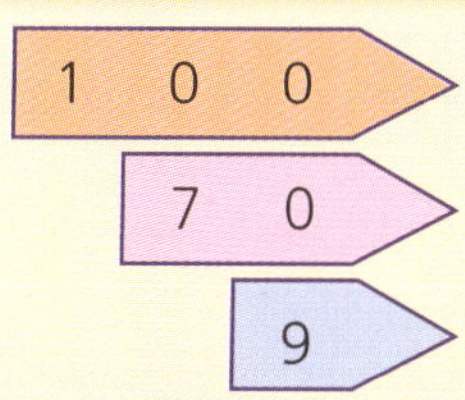

Arrow cards

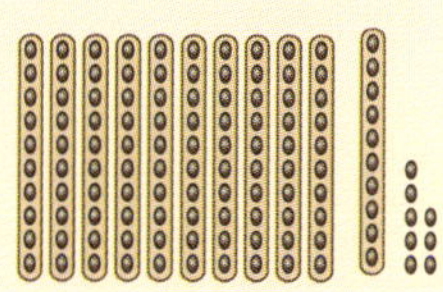

Bean sticks

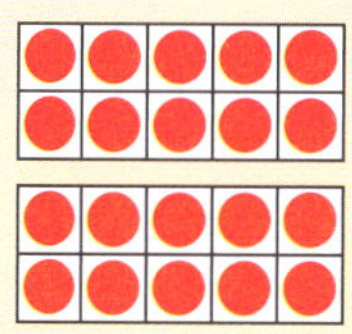

Ten frames

- Use some concrete materials to model these numbers.
  - 184
  - 267
  - 399
  - 701
  - 625
  - 111
  - 430
- Explain your answers to a friend.

9C Features of 2D Shapes

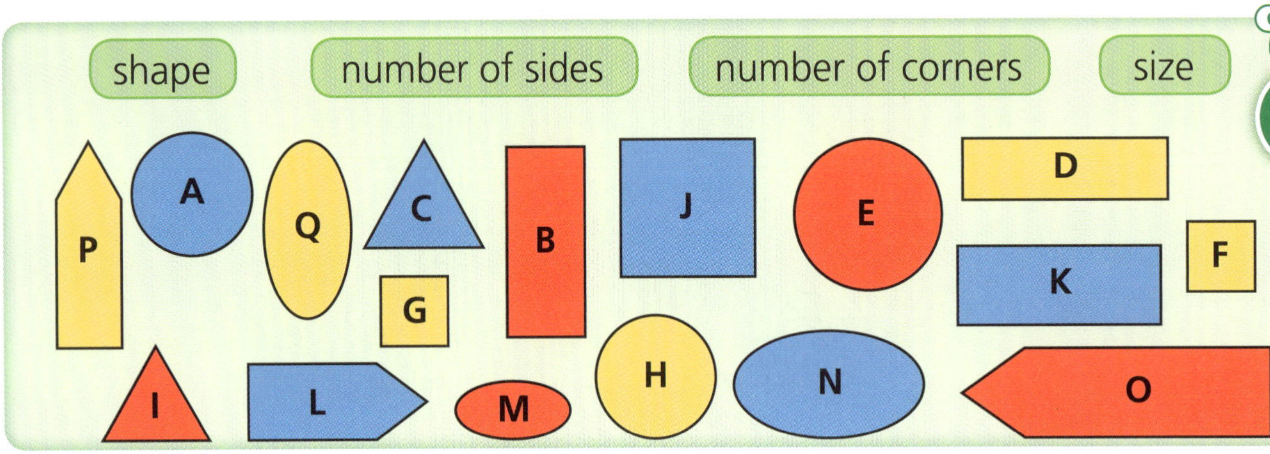

1 Look at these blocks.

a Which have the same shape as E?

b Which have the same shape as K?

c Which has the same shape as I?

d Which have the same shape as G?

e Which has the same shape and size as B?

f Which has the same shape and size as A?

g Which has the same shape and size as G?

h Which have the same shape as P?

i Which have the same shape as Q?

j Which have four corners?

2 What is the name of:

a shape A?

b shape B?

c shape C?

d shape F?

e shape N?

f shape L?

9D Looking at 3D Objects

CONCEPT

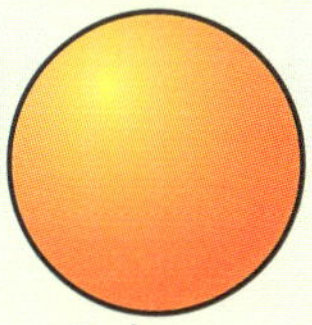

Sphere

One curved surface, no edges

A sphere can roll but not slide.

1 Write some examples of spheres.

2 Choose words from the list that describe each object.

straight
rounded
smooth
flat
pointed
curved

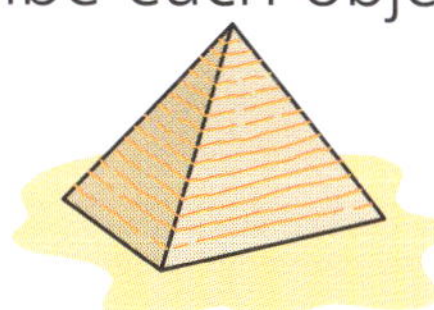

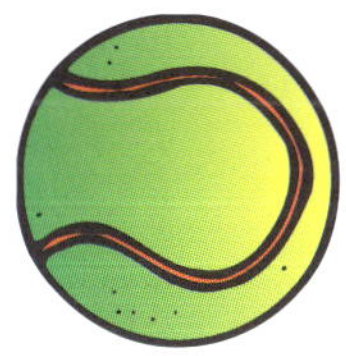

ACTIVITY

3 **a** Draw a sphere. **b** Draw a 3D object that has a point.

10A Problem Solving

Each bag has 5 apples.

To find how many apples are in 4 bags, we count the 5 apples four times.

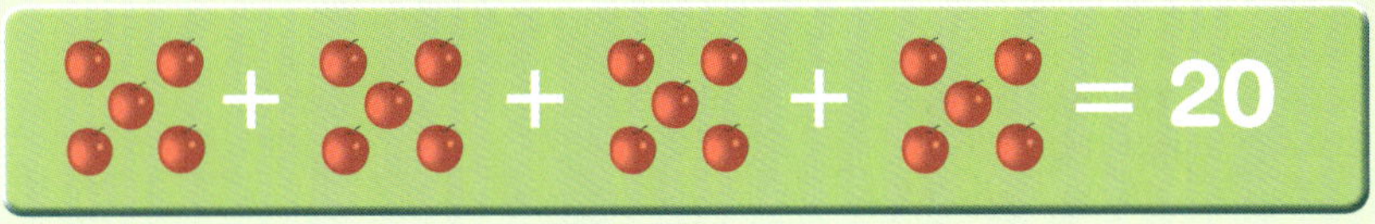

1. Use the picture above to find how many apples are in:

a 2 bags ☐ **b** 3 bags ☐ **c** 4 bags ☐

2. 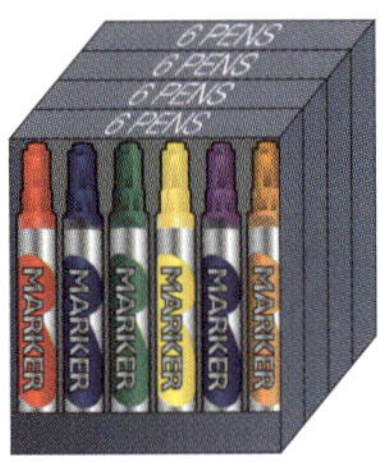

In each packet there are 6 pens.
How many pens are in:

a 2 packets? ☐ **b** 3 packets? ☐

3. 

In each pod there are 4 peas.
How many peas are in:

a 2 pods? ☐ **b** 3 pods? ☐
c 4 pods? ☐ **d** 5 pods? ☐

4. 

In each can there are 3 tennis balls.
How many balls are in:

a 2 cans? ☐ **b** 3 cans? ☐
c 4 cans? ☐ **d** 5 cans? ☐

10B Multiplication Sign

CONCEPT

3 rows of 4 = 12

3 × 4 = 12

1 Complete:

a

2 groups of 3 = □

2 × 3 = □

b

□ groups of □ = □

□ × □ = □

c

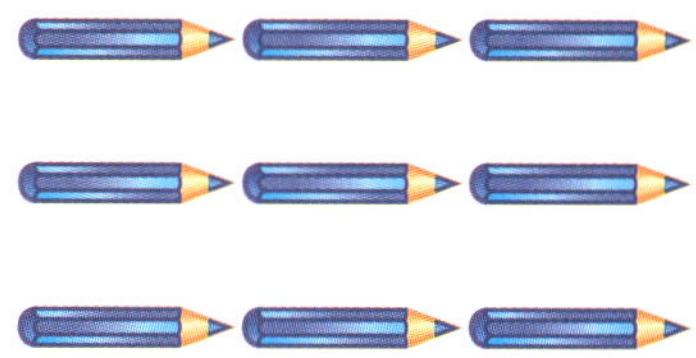

□ groups of □ = □

□ × □ = □

d

□ × □ = □

e

□ × □ = □

f

□ × □ = □

g

□ × □ = □

h

□ × □ = □

i

□ × □ = □

2 Use counters to make groups to answer these questions.

a 4 × 2 = □ **b** 3 × 5 = □ **c** 2 × 4 = □

d 5 × 4 = □ **e** 4 × 4 = □ **f** 5 × 5 = □

10C Chance

ACTIVITY

Use these words to talk about the picture.

likely | unlikely | impossible | possible

Circle the things that might happen this week.

Area is the amount of surface.

The area of the blue rectangle is smaller than the area of the orange rectangle.

1 Use the given unit to find each area. Complete the table.

Area	Unit	Number of Units	
		Estimate	Measure
		☐ books	☐ books
		☐ sheets	☐ sheets

2 Use a rock and then a 50c coin to cover the green square.

a Number of times the rock was used is ☐.

b Number of times the coin was used is ☐.

c Did you cover the whole square? ☐

11A Subtraction to 20

1 Complete these by counting on or counting back.

a

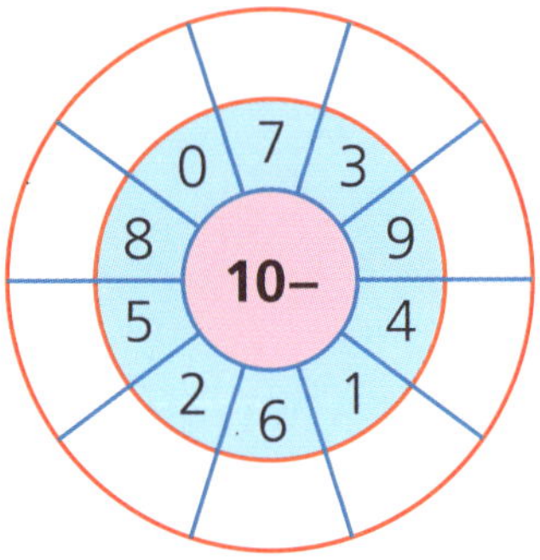

b

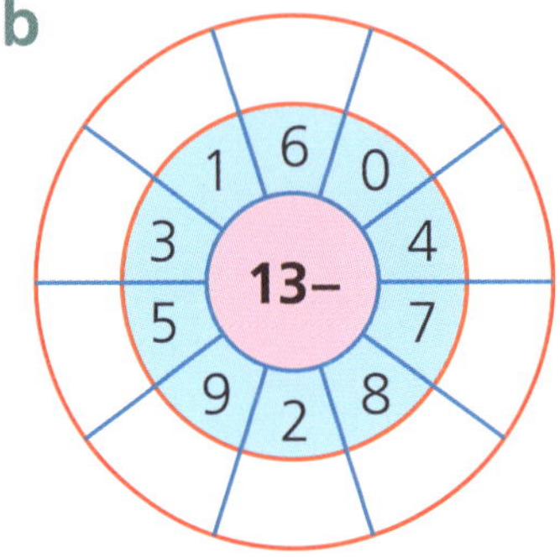

c

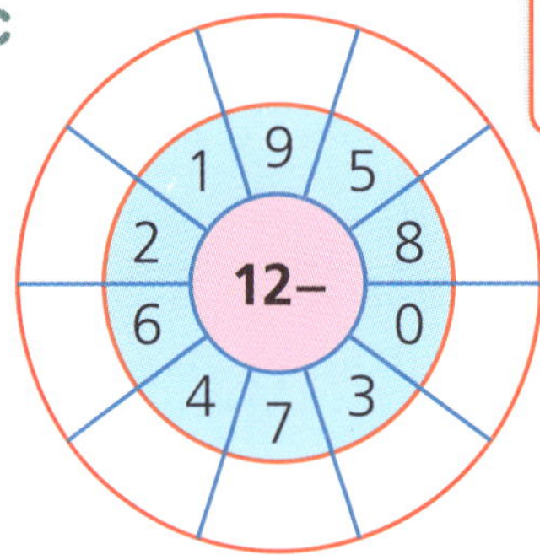

2 Colour the boxes. 10 = red, 9 = blue, 8 = green, 7 = yellow

14 – 4	13 – 4	9 – 2	13 – 6	11 – 4	16 – 7	20 – 10
12 – 3	15 – 5	14 – 5	12 – 5	15 – 6	19 – 9	17 – 8
11 – 3	12 – 4	16 – 6	19 – 10	18 – 8	14 – 6	16 – 8
8 – 1	15 – 7	18 – 9	17 – 7	11 – 2	10 – 3	13 – 5

3 Use this code to find the message.

A	E	H	I	K	L	M	S	T
8	6	2	3	9	4	5	7	10

Message:

10–7	12–8	9–6	11–2	10–4	8–3	9–1	10–0	7–5	9–2

Make a message of your own using the code.

 • *AUSTRALIAN SIGNPOST MATHS 2* • ISBN 978 1 4886 2181 9

11B Number Lines

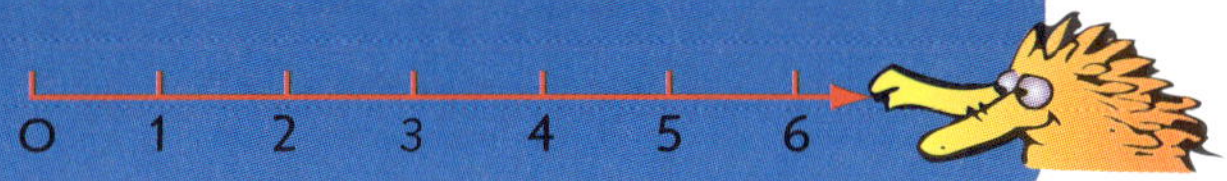

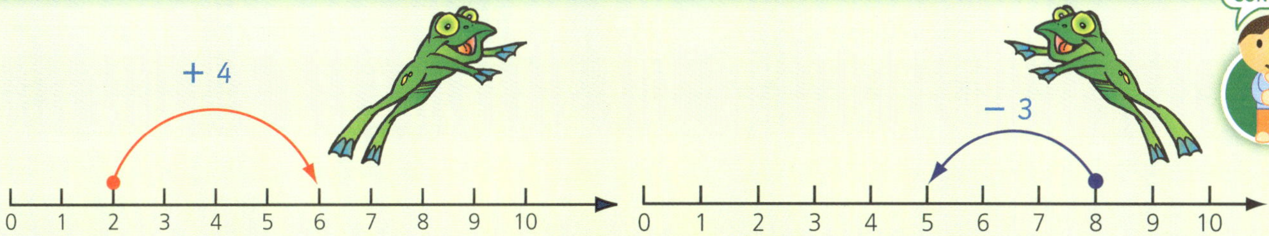

2 + 4 = 6   8 − 3 = 5

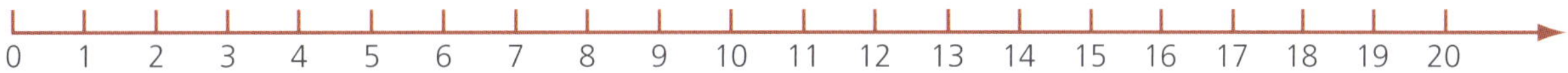

1 Use the number line to complete these.

a Go to 6. Jump 3 places right. 6 + 3 = ☐

b Go to 12. Jump 4 places right. ☐ + ☐ = ☐

c Go to 15. Jump 6 places left. 15 − 6 = ☐

d Go to 17. Jump 8 places left. ☐ − ☐ = ☐

2

a 6 − 3 = ☐   b 13 − 6 = ☐   c 9 − 5 = ☐

d 14 − 9 = ☐   e 16 − 8 = ☐   f 15 − 9 = ☐

3 Fill the empty boxes using subtraction facts.

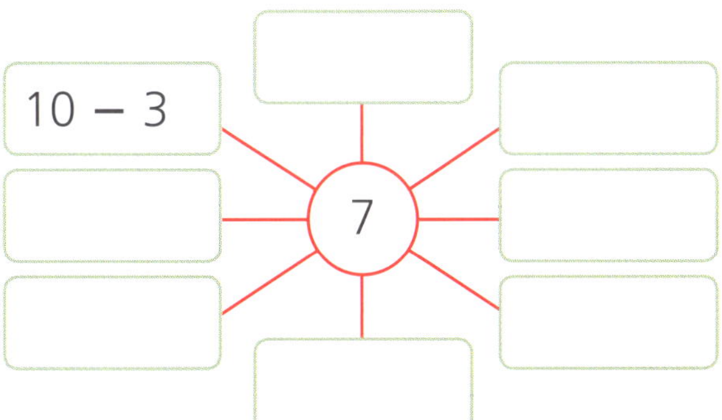

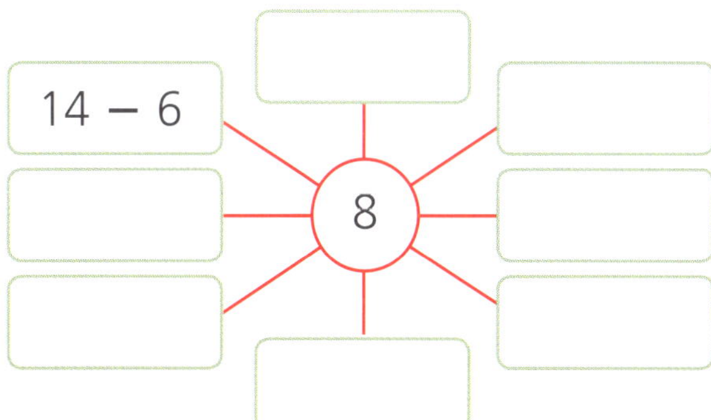

 • *AUSTRALIAN SIGNPOST MATHS 2* • ISBN 978 1 4886 2181 9

11C Number Lines

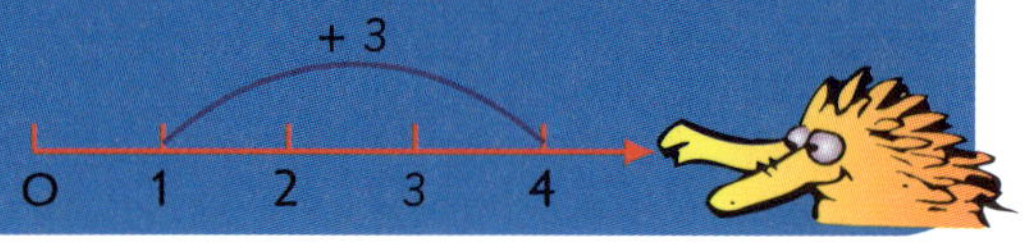

Number line	Number sentence
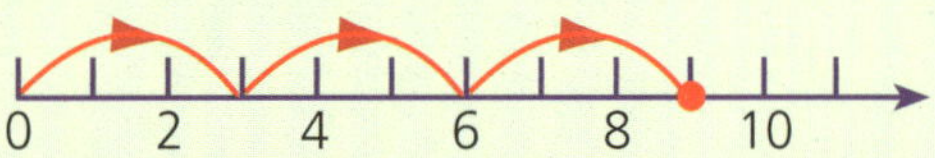	3 + 3 + 3 = 9 or 3 × 3 = 9

1 Write a number sentence to match each number line.

Number sentence

a

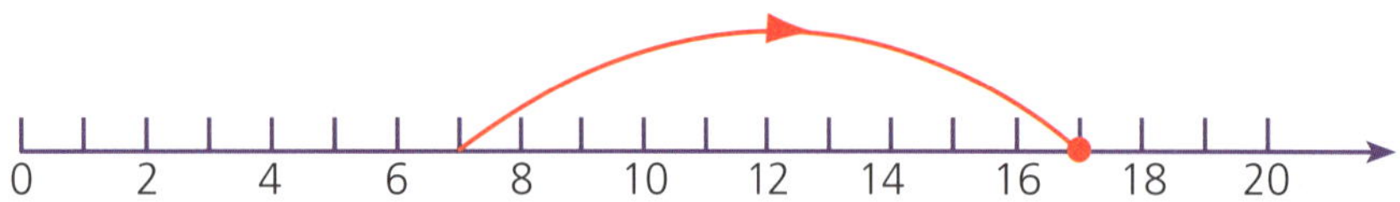

b

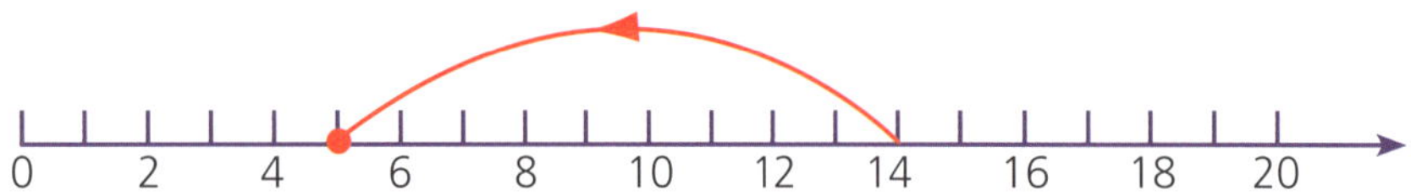

c

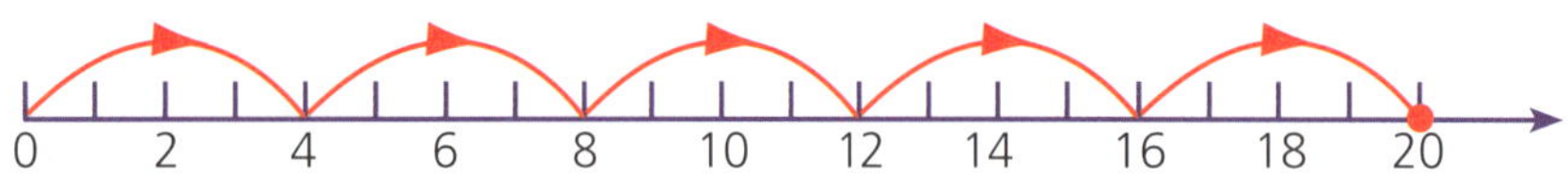

d

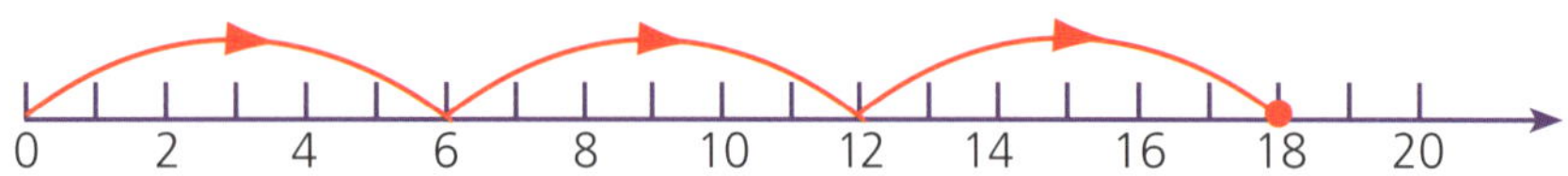

e

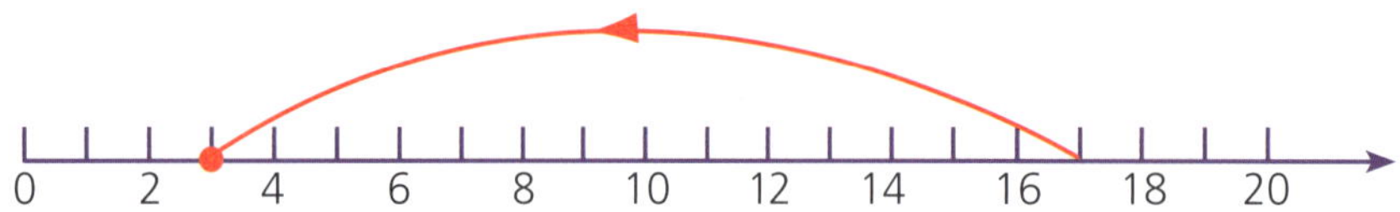

f

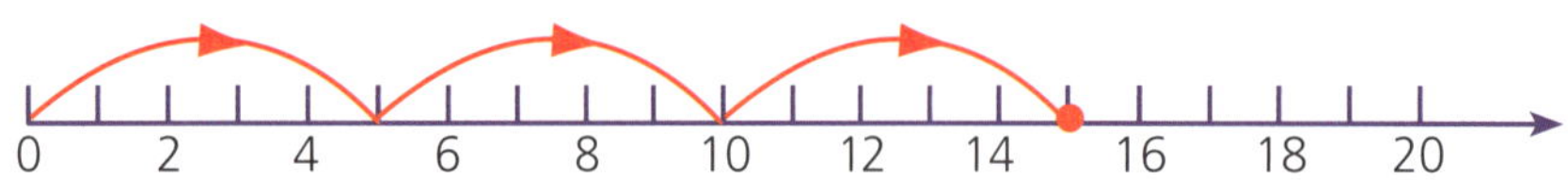

2 Make up a number line and number sentence of your own.

 • *AUSTRALIAN SIGNPOST MATHS 2* • ISBN 978 1 4886 2181 9

11D Cones and Cylinders

Cone

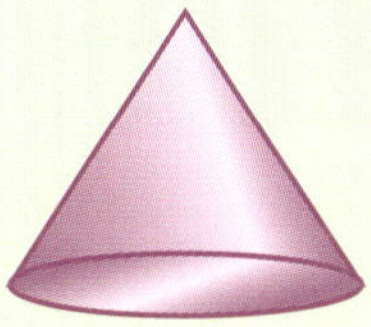

One curved surface,
one circular face

Cylinder

Equal circles are joined
by a curved surface.

1 Write **cone** or **cylinder** under each object.

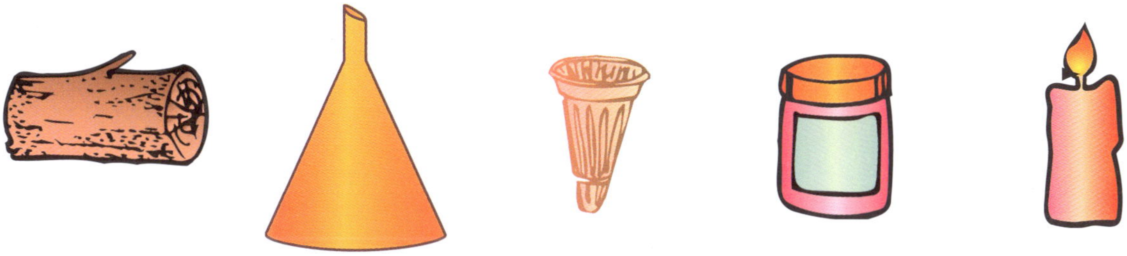

2 Use words like **smooth, corner, edge, face** (flat surface) and **curved surface** to describe each 3D object.

a

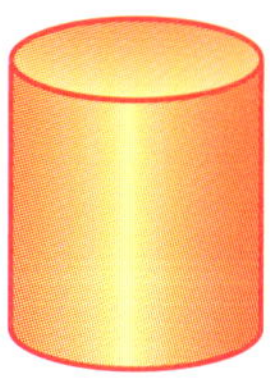

b

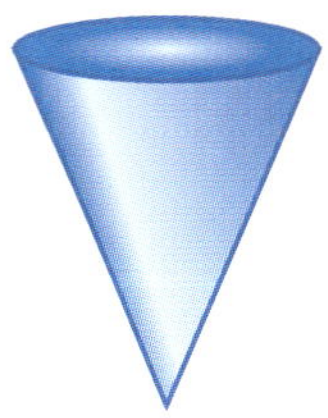

3 Complete the pattern of cones and cylinders.

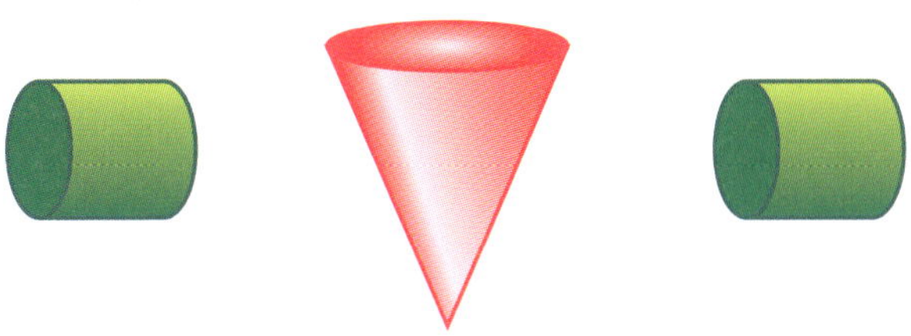

4 Use Plasticine to model cones and cylinders.

12A Fair Shares

1 Mr Lee had 4 sandwiches. He gave each of his 4 children 1 sandwich.

a Did each child get a fair share?

b In the picture, how many cakes does Mr Lee have?

c If the cakes were shared fairly, how many cakes would each child get?

8 shared among 4

2 **a**

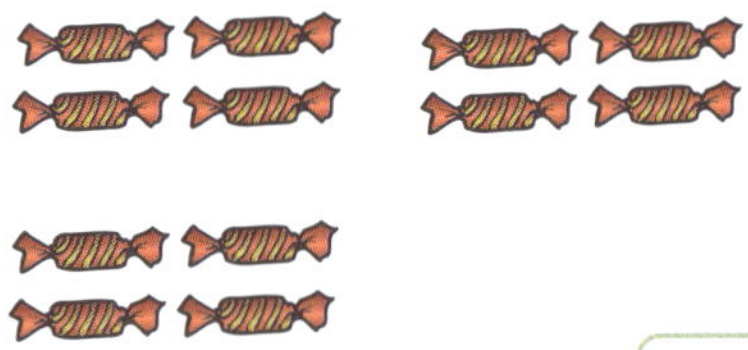

Are these fair shares?

How many lollies are in each share?

b

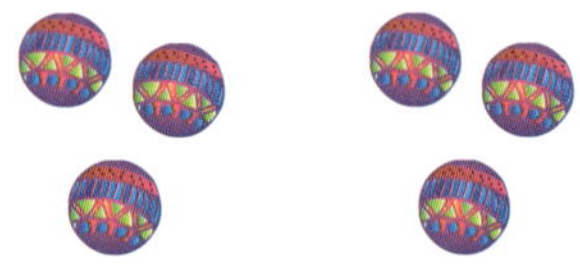

Are these fair shares?

How many balls are in each share?

c

Are these fair shares?

Why?

d

Are these fair shares?

Why?

e Draw more cakes to make equal shares.

How many are in each share now?

Practise sharing counters and blocks with others.

12B Division

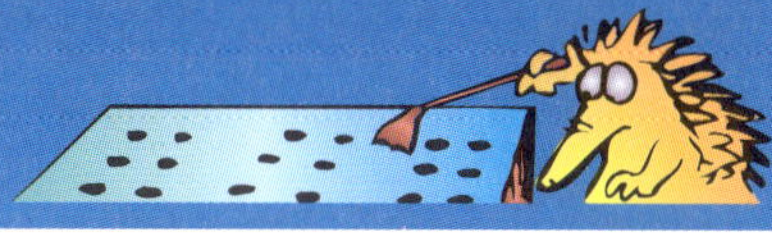

1. Share 6 counters among the boxes.

Share one counter at a time.

2. Share 12 ice creams among 4 children.

Carlos	Lee	Alex	Mina

3. Colour the sets of objects that show a fair share.

a

b

c

d
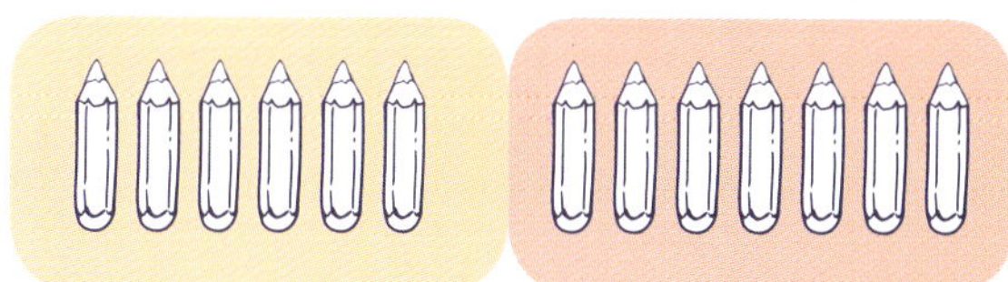

4. I see 24 spider legs.

How many spiders are there?

Use counters.

12C Halves and Quarters

1 Is each group divided into halves or into quarters?

a 

halves

b 

c 

d 

e 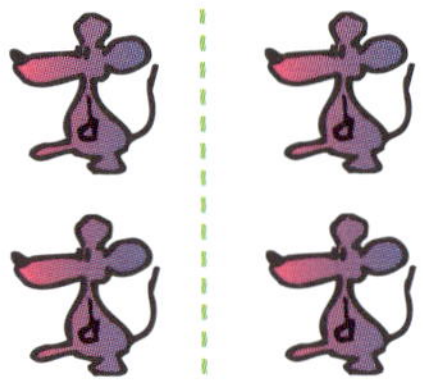

f 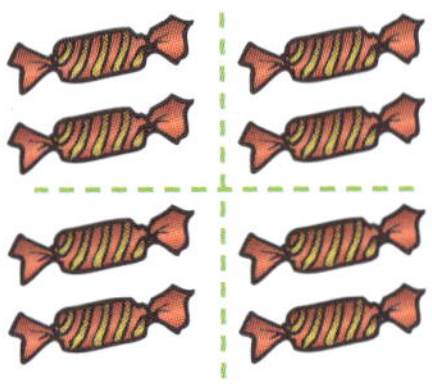

g 

2 Circle part of the group to match the fraction given.

a 

$\frac{1}{2}$

b 

$\frac{1}{4}$

c 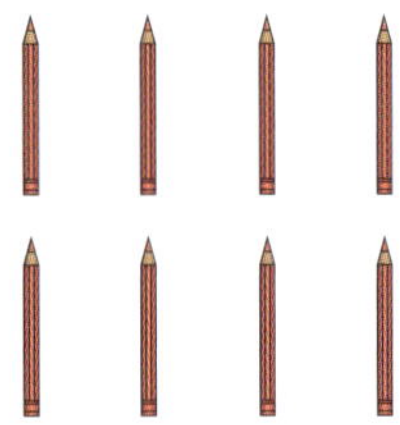

$\frac{1}{2}$

d 

$\frac{1}{4}$

e 

$\frac{1}{4}$

f 

$\frac{1}{4}$

g 

$\frac{1}{2}$

3 Divide each shape into quarters. Colour the fraction shown.

a 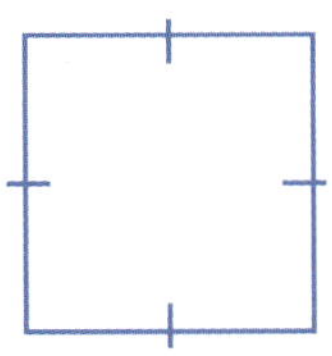 $\frac{1}{2}$

b 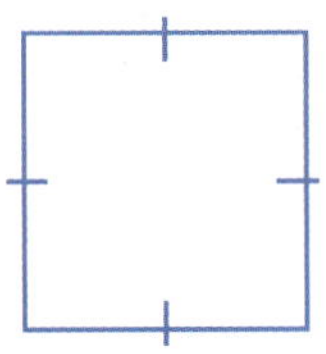 $\frac{1}{4}$

c 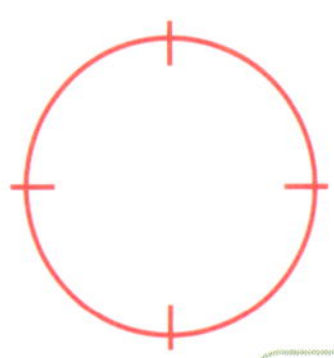 $\frac{1}{2}$

d 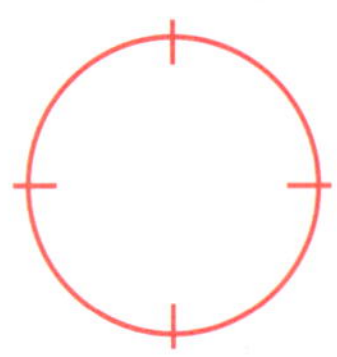 $\frac{1}{4}$

Which fraction is the bigger piece: $\frac{1}{2}$ or $\frac{1}{4}$?

 • *AUSTRALIAN SIGNPOST MATHS 2* • ISBN 978 1 4886 2181 9

12D Halves and Quarters

1 a Fold a square of paper in half.

b Unfold the paper and colour half.

c Fold it back in half.

d Fold it in half again.

e What did you discover?

- Folding paper into halves results in more / fewer parts than if the paper is folded into quarters.

2 Colour part of each group to match the given fraction.

a 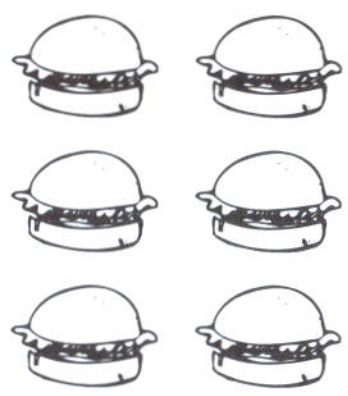 $\frac{1}{2}$

b 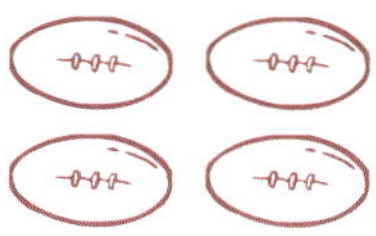 $\frac{1}{4}$

c 

$\frac{1}{2}$

d  $\frac{1}{4}$

3 What part of each group is coloured?

a 

b 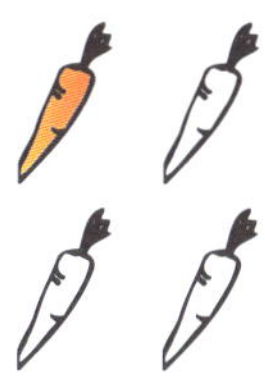

c 

d 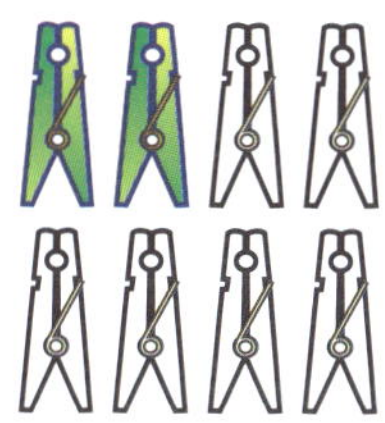

4 What part of each shape is coloured?

a 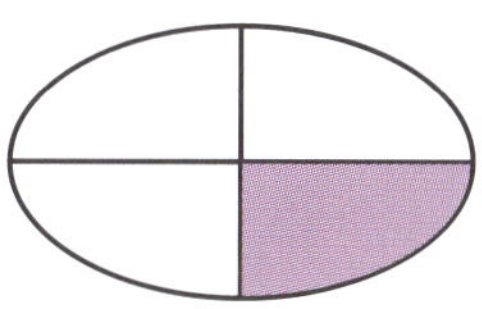

b 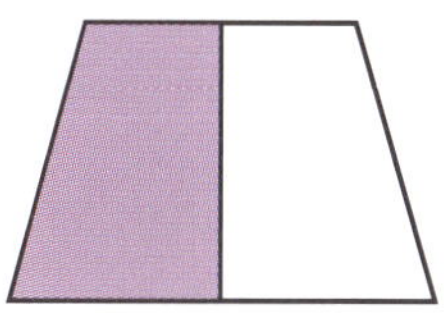

c 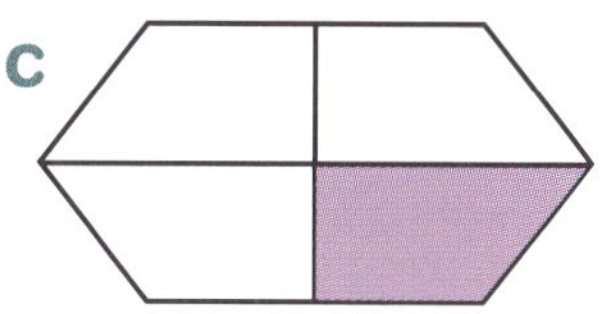

d 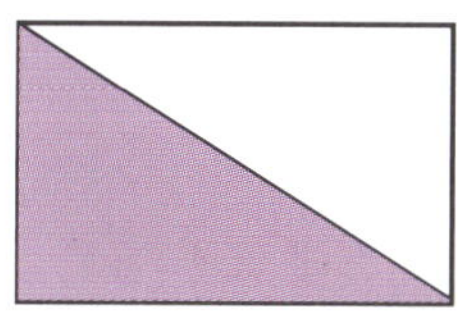

13A Linking Addition and Subtraction

1

10 – 6 = ☐	15 – 7 = ☐	17 – 10 = ☐	13 – 5 = ☐
6 + 4 = ☐	7 + 8 = ☐	10 + 7 = ☐	5 + 8 = ☐

2 Use the pictures to help fill in the missing numerals.

a

5 + 3 = ☐

3 + 5 = ☐

8 – 5 = ☐

8 – 3 = ☐

b

☐ + ☐ = ☐

☐ + ☐ = ☐

10 – ☐ = ☐

10 – ☐ = ☐

c

☐ + ☐ = ☐

☐ + ☐ = ☐

☐ – ☐ = ☐

☐ – ☐ = ☐

d

☐ + ☐ = ☐

☐ + ☐ = ☐

☐ – ☐ = ☐

☐ – ☐ = ☐

e

☐ + ☐ = ☐

☐ + ☐ = ☐

☐ – ☐ = ☐

☐ – ☐ = ☐

3 Write two addition and two subtraction facts for this picture.

☐ + ☐ = ☐

☐ + ☐ = ☐

☐ – ☐ = ☐

☐ – ☐ = ☐

 AUSTRALIAN SIGNPOST MATHS 2 • ISBN 978 1 4886 2181 9

13B Linking Addition and Subtraction

6 + 4 = 10

so 10 − 6 = 4 and 10 − 4 = 6

12 + 8 = 20 → 20 − 8 = 12, 20 − 12 = 8

1 Write two linking subtraction number sentences for each addition.

a 8 + 7 = 15

___ − ___ = ___

___ − ___ = ___

b 6 + 8 = 14

___ − ___ = ___

___ − ___ = ___

c 14 + 6 = 20

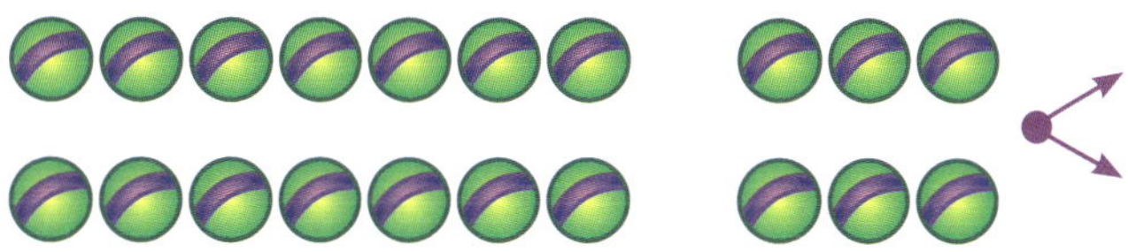

___ − ___ = ___

___ − ___ = ___

d 7 + 5 = 12

___ − ___ = ___

___ − ___ = ___

e 22 + 9 = 31

___ − ___ = ___

___ − ___ = ___

f 17 + 9 = 26

___ − ___ = ___

___ − ___ = ___

Make linked number sentences of your own.

☐ + ☐ = ☐   ☐ − ☐ = ☐   ☐ − ☐ = ☐

 • *AUSTRALIAN SIGNPOST MATHS 2* • ISBN 978 1 4886 2181 9

13C Left and Right

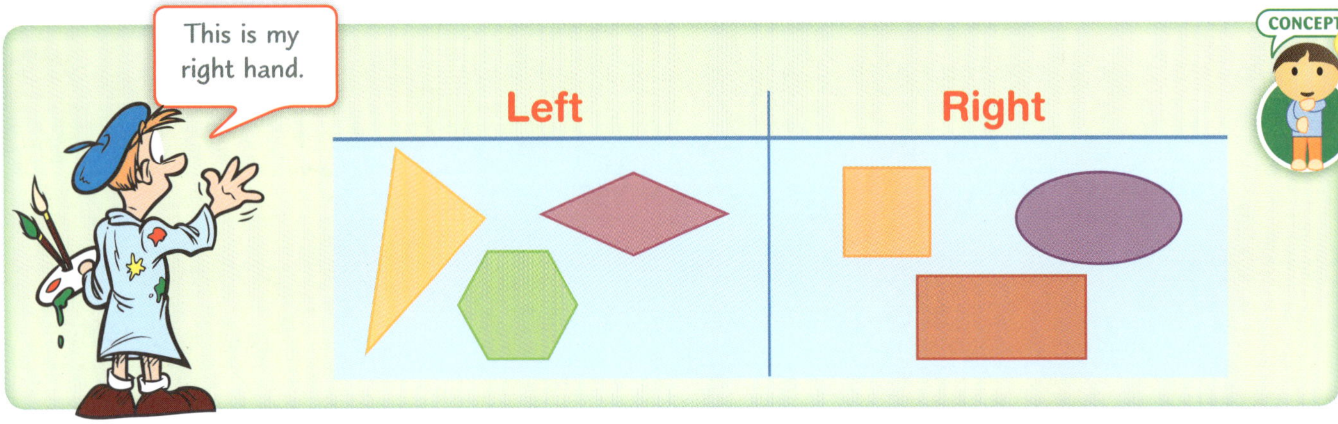

1. **a** What shapes are on the left?

   b What shapes are on the right?

2. What is:

   a to the left of the plate?

   b to the right of the pen?

   c in front of the plate?

   d behind the plate?

   e to the left of the jug?

   f in front of the table?

   g to the right of the jug?

   h to the right of the spoon?

   i behind the pen?

   j to the left of the apple?

hat	jug	apple
spoon	plate	cup
pen	chair	table

INVESTIGATION

Ask classmates or people you know whether they are right-handed or left-handed. Keep a count.

Number left-handed: ☐ Number right-handed: ☐

 • *AUSTRALIAN SIGNPOST MATHS 2* • ISBN 978 1 4886 2181 9

13D The Trapezium and Kite

1 Name the shapes used.

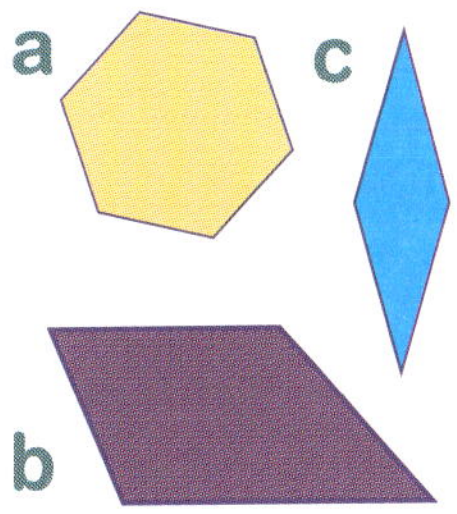

a	
b	
c	

circle rhombus triangle square rectangle hexagon trapezium

CONCEPT

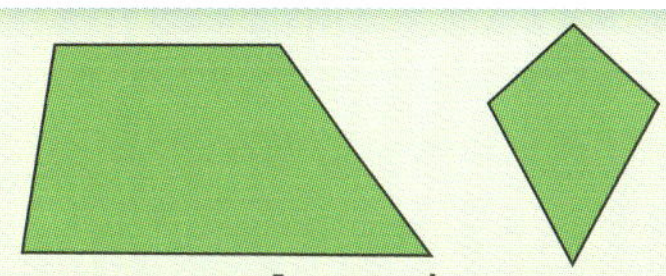

A **trapezium** has a pair of parallel sides.

A **kite** has two pairs of equal sides.

2 Trace around the shape. Write its name.

3 Colour the trapeziums red and the kites blue.

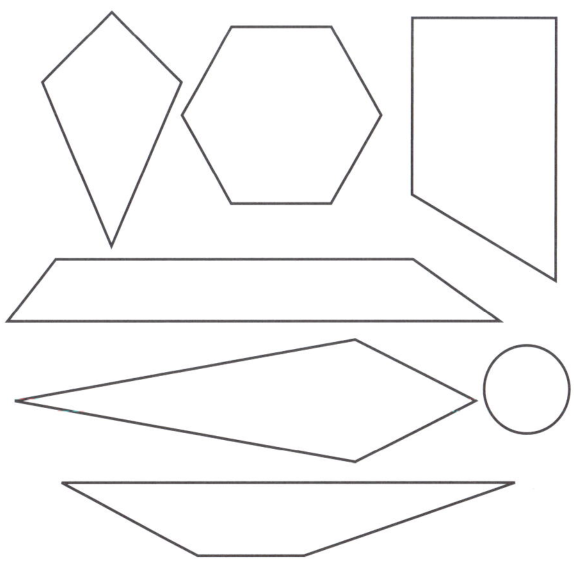

4 Draw a trapezium and a kite.

5 a How many sides has a trapezium?

b How many corners has a trapezium?

c Does a trapezium have any parallel sides?

 • *AUSTRALIAN SIGNPOST MATHS 2* • ISBN 978 1 4886 2181 9

14A Sharing

CONCEPT

When we share, each person is given the same number of items.

1. Circle each share, then answer the questions.

a Shared among 4

How many stars? ☐

How many shares? ☐

How many stars in each share? ☐

b Shared among 4

How many caps? ☐

How many shares? ☐

How many caps in each share? ☐

c Shared between 2

How many pears? ☐

How many shares? ☐

How many pears in each share? ☐

d Shared among 4

How many fish? ☐

How many shares? ☐

How many fish in each share? ☐

e Shared between 2

How many stars? ☐

How many shares? ☐

How many stars in each share? ☐

f Shared among 4

How many cups? ☐

How many shares? ☐

How many cups in each share? ☐

INVESTIGATION

2. Use 18 counters. What is one share if they are shared:

a among 3? ☐ counters **b** among 6? ☐ counters

3. Use 20 counters. What is one share if they are shared:

a among 5? ☐ counters **b** among 4? ☐ counters

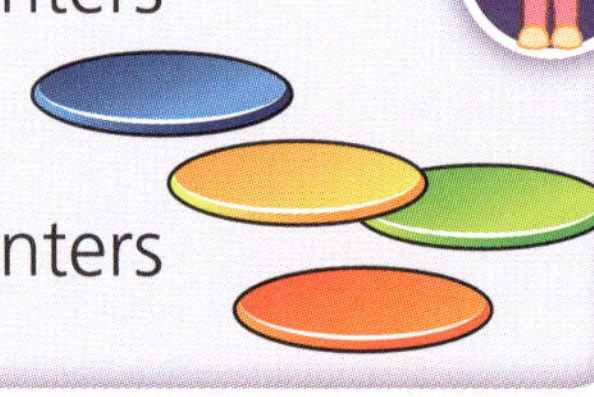

 • *AUSTRALIAN SIGNPOST MATHS 2* • ISBN 978 1 4886 2181 9

14B Using Groups

CONCEPT

We can use counters or blocks to model problems.

12 apples. Put 4 in each box.
How many boxes do we need?

Answer: We would need 3 boxes. We could circle groups of 4.

1 Draw circles to show these groups and rows.

a groups of 3 frogs

How many frogs? ☐

How many groups? ☐

How many frogs in each group? ☐

b rows of 4 jugs

How many jugs? ☐

How many rows? ☐

How many jugs in each row? ☐

c groups of 8 fish

How many fish? ☐

How many groups? ☐

How many fish in each group? ☐

d groups of 2 birds

How many birds? ☐

How many groups? ☐

How many birds in each group? ☐

e groups of 5 ducks

How many ducks? ☐

How many groups? ☐

How many ducks in each group? ☐

f rows of 7 pins

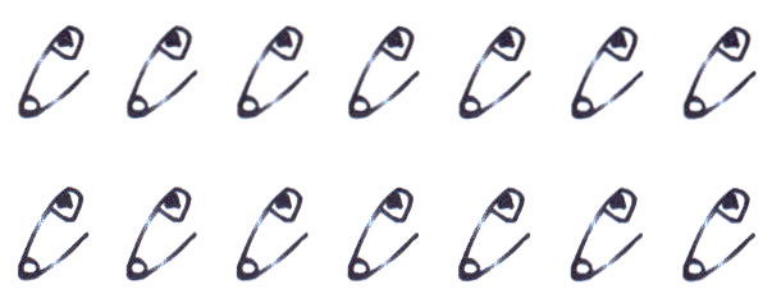

How many pins? ☐

How many rows? ☐

How many pins in each row? ☐

INVESTIGATION

2 Use 24 counters. How many girls can be given:

a 2 counters? ☐ **b** 3 counters? ☐ **c** 4 counters? ☐

d 6 counters? ☐ **e** 8 counters? ☐ **f** 1 counter? ☐

14C Digital Time

1 Write two ways we can say the time shown.

a 

or

b 

or

c 

or

d 

or

e 

or

f 

or

2 Write these as digital times.

a 

b 

 ISBN 978 1 4886 2181 9

14D Lists, Graphs and Tables

We can show data using objects, pictures, words or symbols.

Table

Pet	Number
Cat	2
Dog	5
Rat	1

Graph: Class Pets

List

Dog, Cat, Dog, Dog, Rat, Cat, Dog, Dog

1. Use the data above to answer these questions.

   a How many cats? ☐

   b How many more dogs than cats? ☐

2. People chose their favourite animal. Use the list to finish the graph and table.

List

Monkey, Lion, Lion, Monkey, Elephant, Lion, Monkey, Elephant, Elephant, Elephant, Lion, Elephant, Monkey, Elephant, Monkey

Favourite Animals

Animal	Number
Elephant	
Lion	
Monkey	

Discuss the results.

 • *AUSTRALIAN SIGNPOST MATHS 2* • ISBN 978 1 4886 2181 9

15A Numbers to 1000

CONCEPT

To make large numbers, count the place-value blocks as you go.

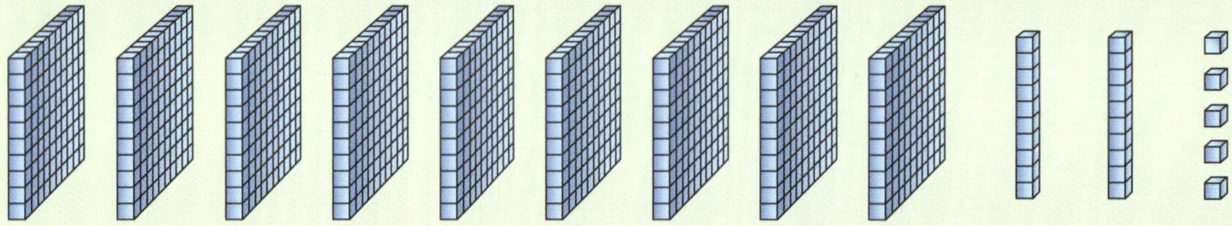

100 ...200 ...300 ...400 ...500 ...600 ...700 ...800 ...900 ...910 ...920 ...925

9 hundreds + 2 tens + 5 ones = 925

Hundreds	Tens	Ones
IIIIIIIII	II	IIIII
9	2	5

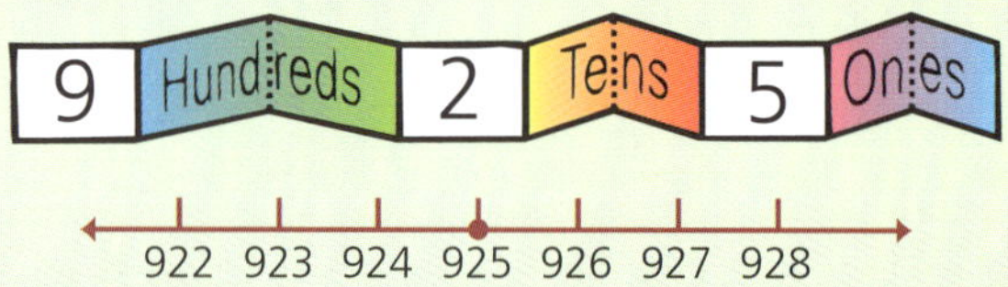

1 Complete each part.

a

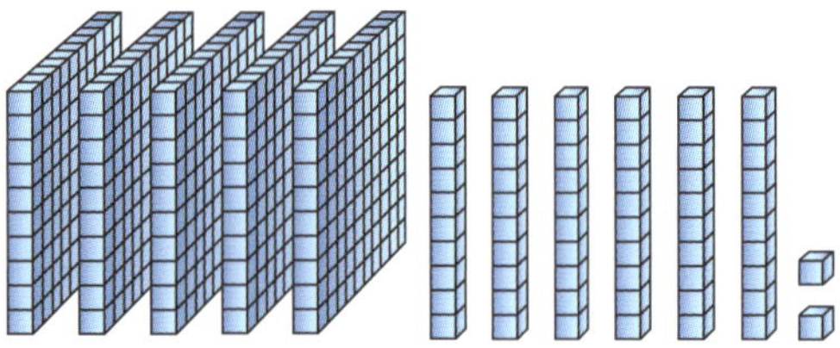

☐ hundreds + ☐ tens + ☐ ones

Hundreds	Tens	Ones
☐	☐	☐

= ☐

b

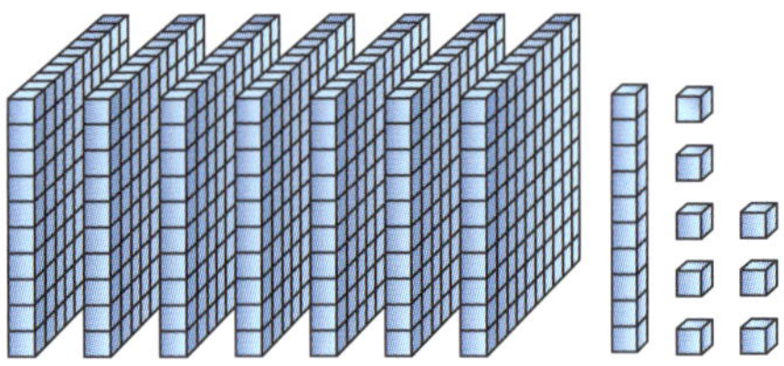

☐ hundreds + ☐ tens + ☐ ones

Hundreds	Tens	Ones
☐	☐	☐

= ☐

c

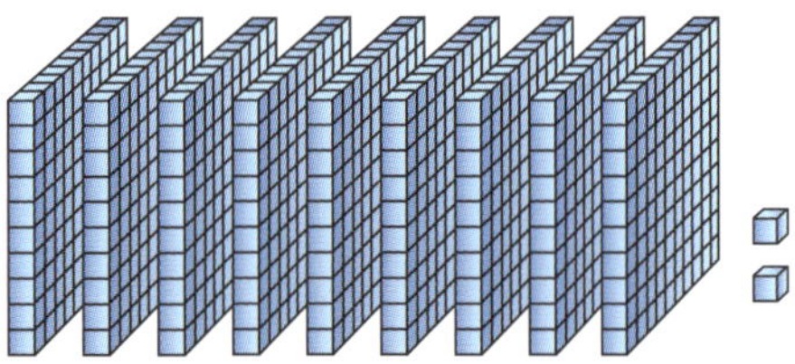

☐ hundreds + ☐ tens + ☐ ones

Hundreds	Tens	Ones
☐	☐	☐

= ☐

ACTIVITY

Model these numbers using linking blocks, bundles of sticks, place-value blocks or bead strings.

Explain your answers to a friend.

- 734
- 906
- 658
- 812
- 543

 • *AUSTRALIAN SIGNPOST MATHS 2* • ISBN 978 1 4886 2181 9

15B Numbers to 1000

1 Complete each part.

a

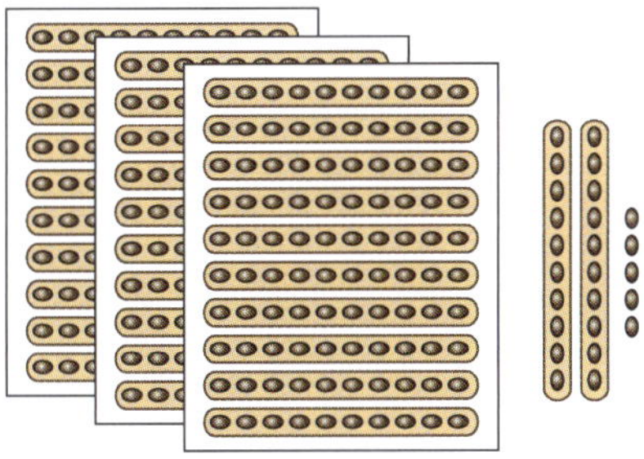

☐ hundreds + ☐ tens + ☐ ones

Hundreds	Tens	Ones
☐	☐	☐

= ☐

b

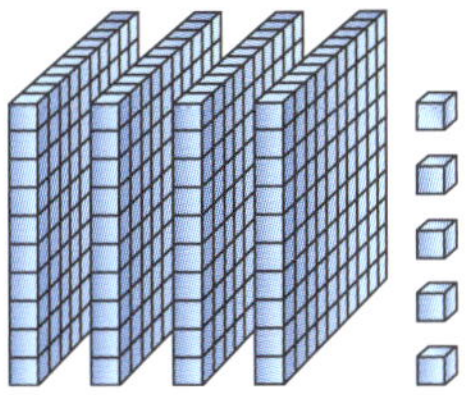

☐ hundreds + ☐ tens + ☐ ones

Hundreds	Tens	Ones
☐	☐	☐

= ☐

c

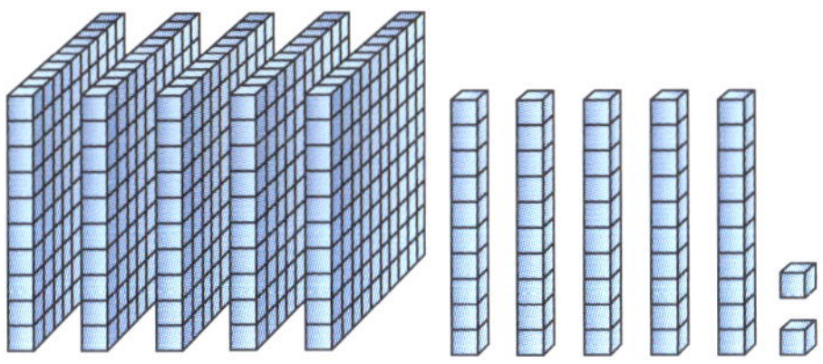

☐ hundreds + ☐ tens + ☐ ones

Hundreds	Tens	Ones
☐	☐	☐

= ☐

d

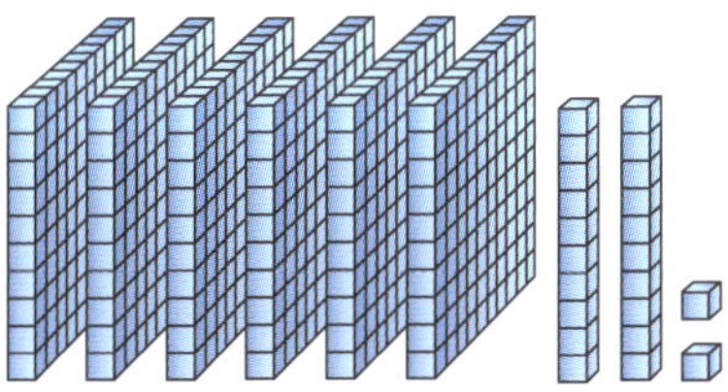

☐ hundreds + ☐ tens + ☐ ones

Hundreds	Tens	Ones
☐	☐	☐

= ☐

e

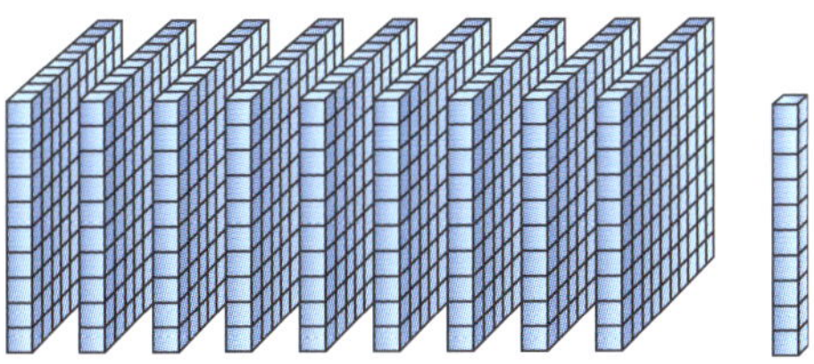

☐ hundreds + ☐ tens + ☐ ones

Hundreds	Tens	Ones
☐	☐	☐

= ☐

- Use place-value blocks to make these numbers.
  - 427
  - 381
  - 836
  - 592
  - 679
  - 764
- Count a collection of 500 bottle lids, buttons or sticks by grouping by tens and hundreds. Practise counting forwards and backwards.

 • *AUSTRALIAN SIGNPOST MATHS 2* • ISBN 978 1 4886 2181 9

15C Reading Graphs

1 This graph shows languages, other than English, spoken by students in Mrs Johnson's class. Use the graph to complete the table.

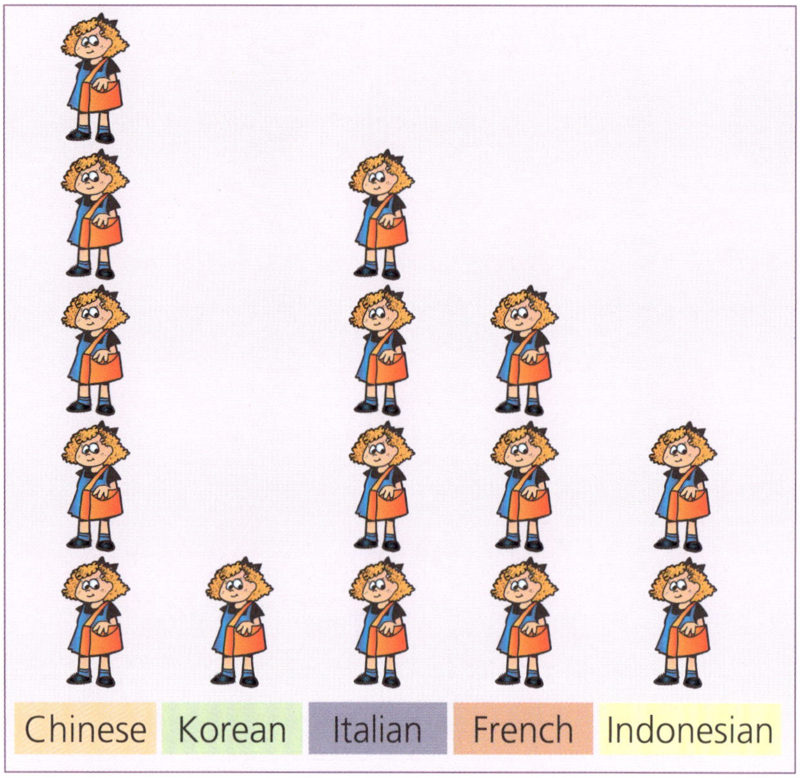

Mrs Johnson's Class	
Language	**Number**
Chinese	
Korean	
Italian	
French	
Indonesian	

Key:  = 1 student

a Which is the biggest group?

b Which is the smallest group?

c How many students speak French or Korean?

d How many students speak Italian or Indonesian?

e How many more students speak Chinese than Korean?

f How many students speak a language other than English?

2 **Favourite Colours**

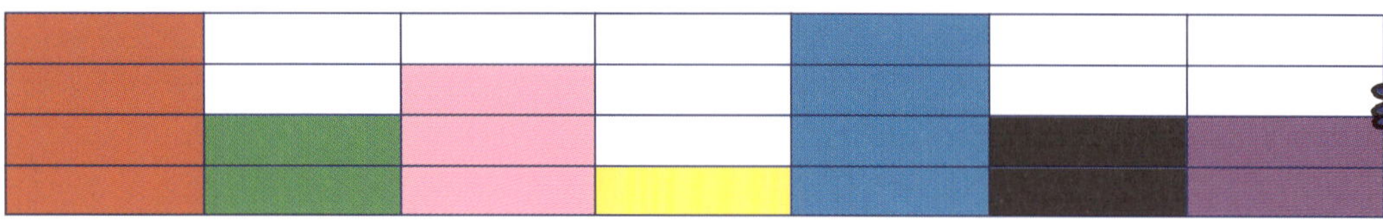

column graph

a Which are the most popular colours?

b How many people told us their favourite colour?

 • *AUSTRALIAN SIGNPOST MATHS 2* • ISBN 978 1 4886 2181 9

15D Using Graphs

❶ How many:

a sandwiches?

b pies?

c apples?

d altogether?

e more pies than apples?

Food in the Classroom

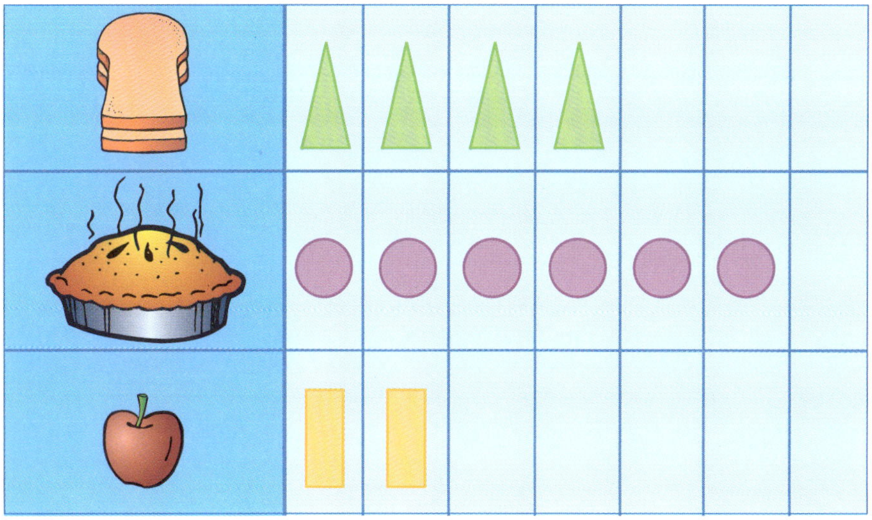

f Complete the table below to show the food in the classroom.

Food in the Classroom	
Sandwiches	
Pies	
Apples	

g What other foods could be in a classroom?

❷ Each girl put a candle above her month of birth to make this graph.

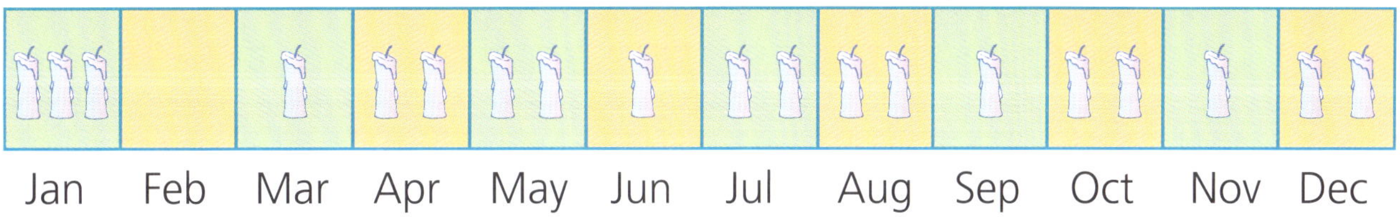

How many girls were born in:

a January?

b November?

c February?

d How many girls were there altogether?

The Colour of Cars

Discuss how your class could find out the most popular colours of the cars passing the school.

 • *AUSTRALIAN SIGNPOST MATHS 2* • ISBN 978 1 4886 2181 9

16A Ordering Numbers

Some numbers are smaller.

Numbers are written in order.

Some numbers are larger.

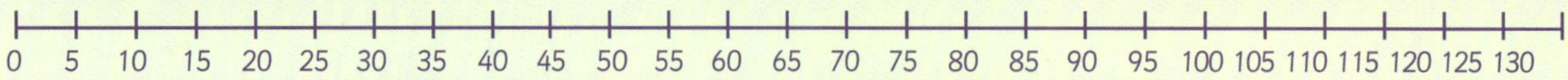

1 Arrange these numbers in order from **smallest** to **largest**.

a 47, 29, 15, 36, 58

b 77, 91, 62, 84, 53, 101

c 83, 105, 130, 127, 99, 116

2 Write the **largest** number possible using these digits.

a 2, 1 **b** 6, 9 **c** 7, 5

3 Write the **smallest** number possible using these digits.

a 1, 2 **b** 8, 7 **c** 4, 8

This is called a number track.

0	1	2	3	4	5	6	7	8	9	10	11	12

, ...

4 Make a number track and cut it into 2s. Mix the pieces up and have your partner time you as you put them in order.

5 Complete this number track.

20	21	22	23	24	25										

6 Starting with zero, complete the track below and write the total of each pair. Can you see a pattern?

0	1	2	3	4	5										

1	5	9					

 • *AUSTRALIAN SIGNPOST MATHS 2* • ISBN 978 1 4886 2181 9

16B Addition by Looking for Tens

1

a
  3
  4
+ 7
☐

b
  8
  5
+ 2
☐

c
  5
  3
+ 5
☐

d
  4
  2
+ 6
☐

e
  9
  9
+ 1
☐

f
  7
  3
+ 8
☐

CONCEPT

  3
  8
  7
  3
+ 2
23

2

a
  9
  1
  2
+ 4
☐

b
  7
  1
  8
+ 3
☐

c
  3
  6
  3
+ 4
☐

d
  6
  5
  4
+ 5
☐

e
  3
  4
  5
+ 7
☐

3

a
  $3
  $2
  $6
+ $4
☐

b
  $2
  $7
  $8
+ $3
☐

c
  $5
  $4
  $1
+ $3
☐

d
  $8
  $3
  $5
+ $7
☐

e
  $9
  $5
  $7
+ $5
☐

4

a 8 cows, 5 sheep, 2 goats.
How many animals? ☐

b 3 cars, 8 trucks, 7 bikes.
What is the total? ☐

5 True or false?

a 1 + 2 + 8 = 2 + 8 + 1 ☐

b 6 + 8 + 4 = 4 + 6 + 8 ☐

c 3 + 5 + 7 = 3 + 7 + 5 ☐

CONCEPT

Fraction of a Whole

CONCEPT

1 What part of each shape is coloured?

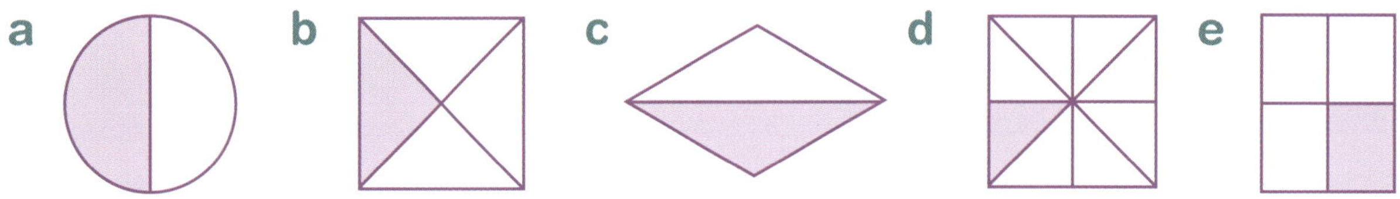

2 Colour part of each shape to match the given fraction.

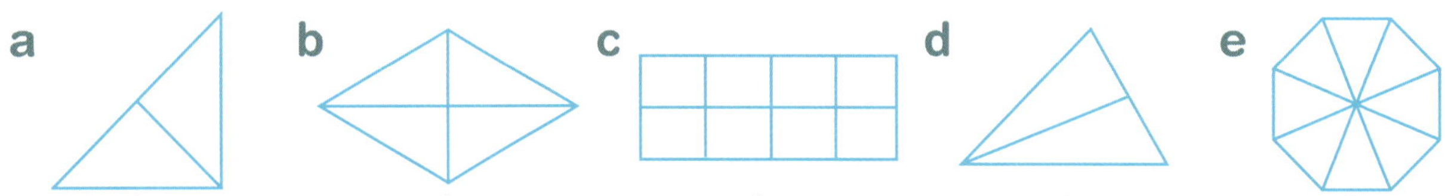

a $\frac{1}{2}$ b $\frac{1}{4}$ c $\frac{1}{8}$ d $\frac{1}{2}$ e $\frac{1}{8}$

3 Colour one half of each shape.

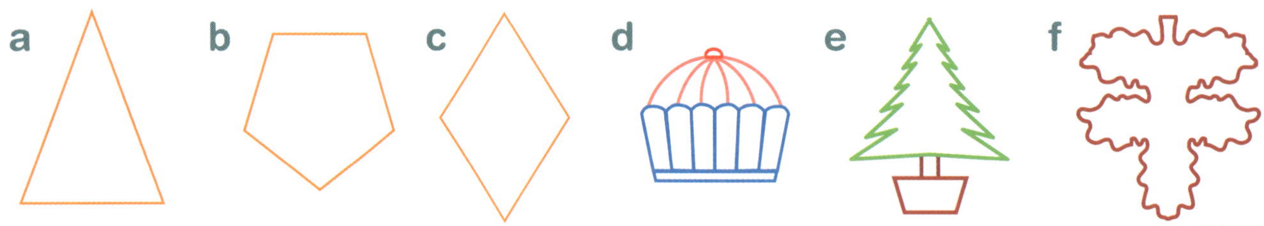

ACTIVITY

Find fractions in the classroom and draw them.

The window has two halves.

16D Fraction of a Group

CONCEPT

1 What part is coloured, one half, one quarter or one eighth?

a  ______

b  ______

c  ______

d  ______

e  ______

f  ______

g  ______

h  ______

2 Colour part of each group to match the given fraction.

a  $\frac{1}{2}$

b  $\frac{1}{4}$

c 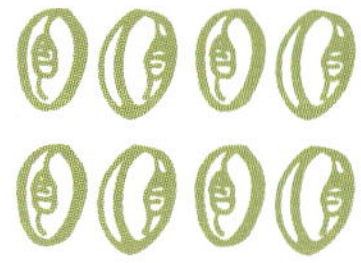 $\frac{1}{8}$

d  $\frac{1}{4}$

e  $\frac{1}{8}$

f  $\frac{1}{4}$

g 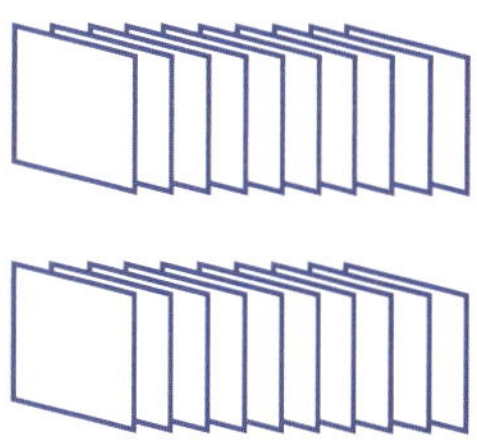 $\frac{1}{2}$

h  $\frac{1}{2}$

3 The fraction coloured is ______.

ACTIVITY

Use counters to model groups split into halves, quarters and eighths.

17A Adding 10s

2 + 10 = 12 4 + 10 = 14

I added each row to my total.

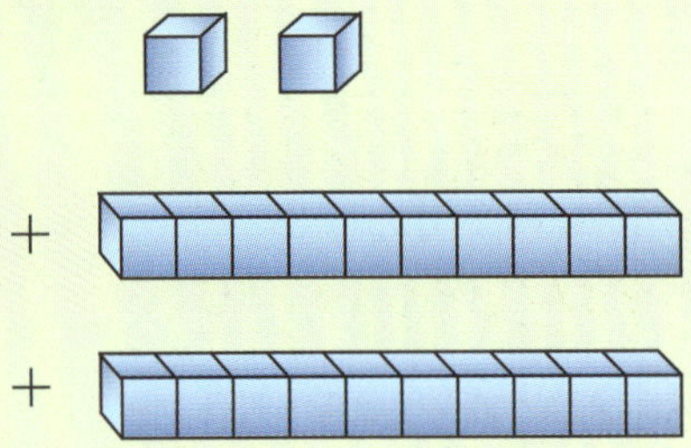

2

12

22

1 Write the totals, adding another ten for each row.

a

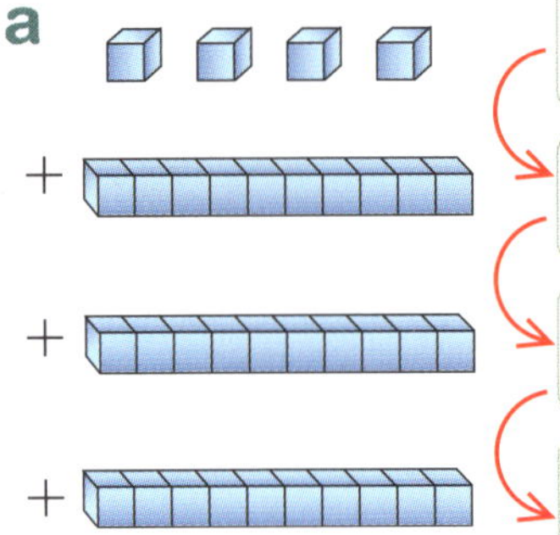

b

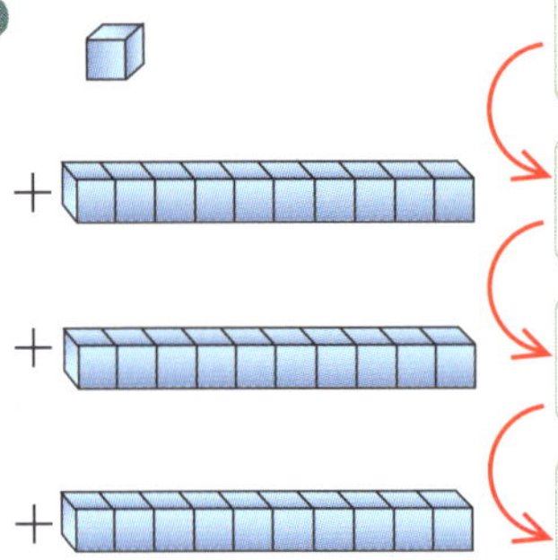

c

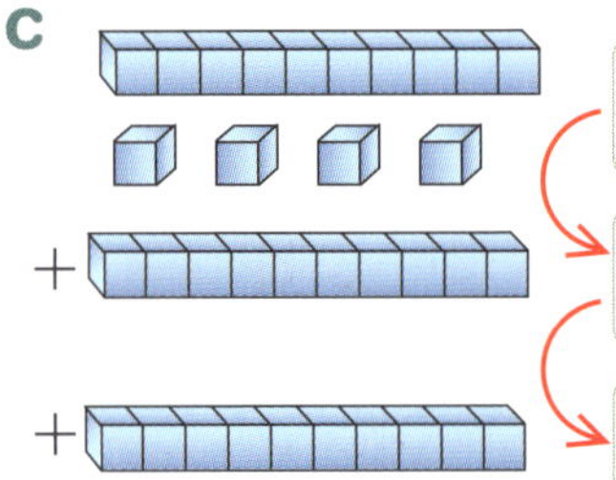

d

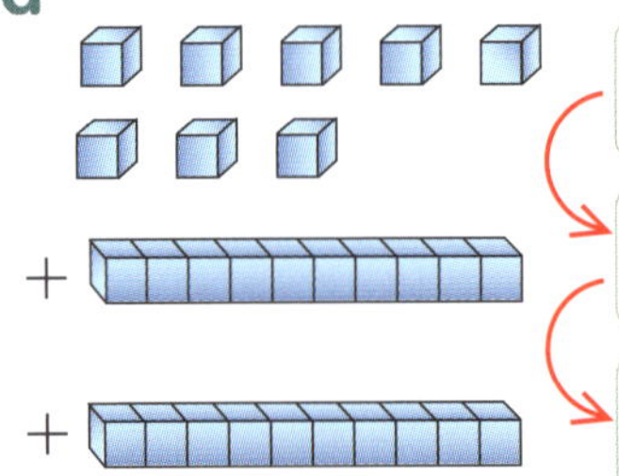

e

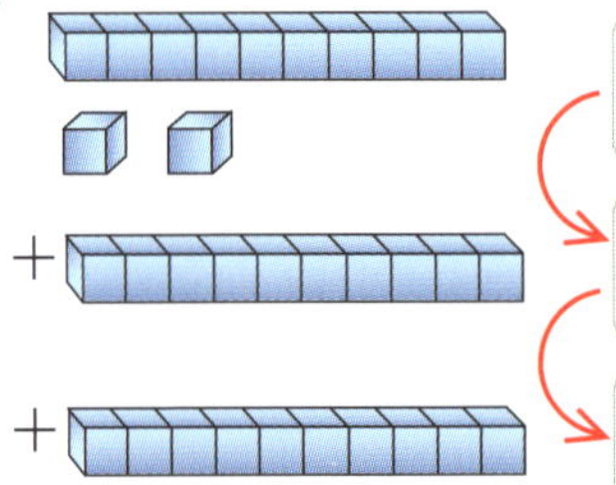

f

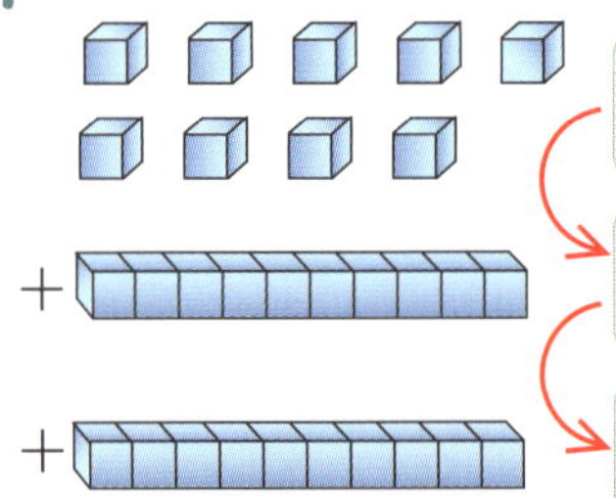

2 Add 10 to each number.

a 3 ☐ **b** 36 ☐ **c** 17 ☐ **d** 52 ☐ **e** 94 ☐

3 Add 20 to each number.

a 1 ☐ **b** 35 ☐ **c** 68 ☐ **d** 19 ☐ **e** 33 ☐

Line up place-value blocks to create your own addition tens pattern.

☐, ☐, ☐, ☐, ☐, ☐

© PEARSON AUSTRALIA 2018 • *AUSTRALIAN SIGNPOST MATHS 2* • ISBN 978 1 4886 2181 9

NUMBER & ALGEBRA

17B Adding and Subtracting 10s

CONCEPT

We can use the columns to add 10 or take away 10.

9, 19, 29 … 19 + 10 = 29

43, 33, 23 … 43 – 10 = 33

1	2	3	4	5	6	7	8	9	10
11	12	13	14	15	16	17	18	19	20
21	22	23	24	25	26	27	28	29	30
31	32	33	34	35	36	37	38	39	40
41	42	43	44	45	46	47	48	49	50

1 Help Banjo find the answers by adding or subtracting 10.

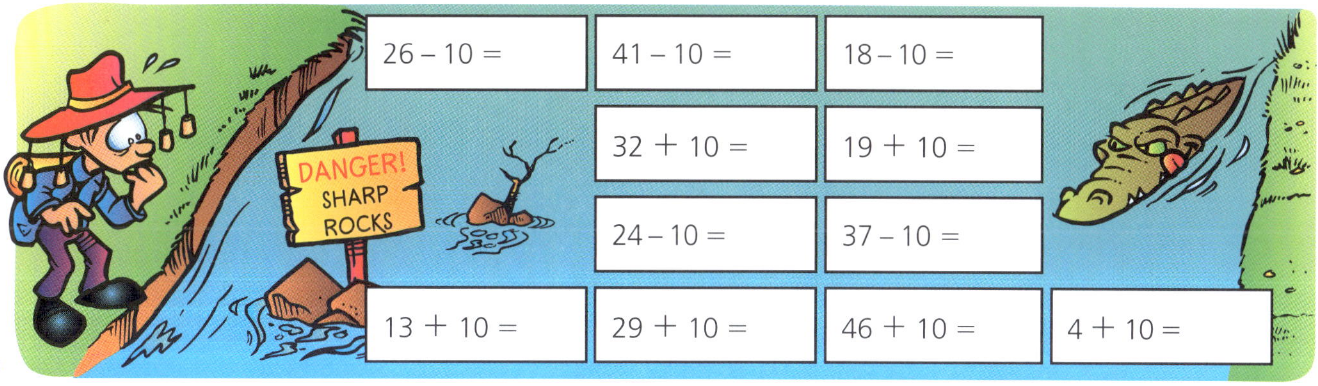

26 – 10 = 41 – 10 = 18 – 10 =

32 + 10 = 19 + 10 =

24 – 10 = 37 – 10 =

13 + 10 = 29 + 10 = 46 + 10 = 4 + 10 =

2 Continue these patterns.

a 15, 25, 35, ☐, ☐, ☐ **b** 59, 49, 39, ☐, ☐, ☐

c 37, 47, 57, ☐, ☐, ☐ **d** 81, 71, 61, ☐, ☐, ☐

3 Use 10s patterns to complete these.

a 2 + 20 = 2 + 10 + 10 2, ☐, ☐ 2 + 20 = ☐

b 27 + 30 = ☐ 27, ☐, ☐, ☐ 27 + 30 = ☐

c 51 – 20 = 51 – 10 – 10 51, ☐, ☐ 51 – 20 = ☐

d 85 – 30 = ☐ 85, ☐, ☐, ☐ 85 – 30 = ☐

e 43 + 40 = ☐ **f** 74 – 40 = ☐ **g** 38 + 40 = ☐

17C Analogue Time

4 o'clock

a quarter past 4

half past 4

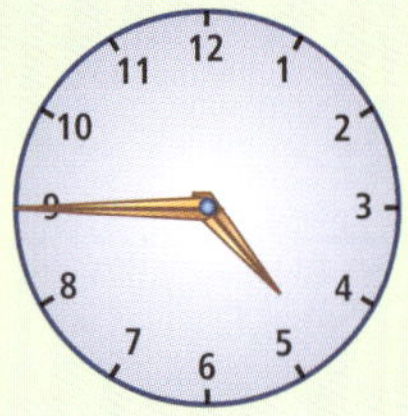

a quarter to 5

1 Write the time shown on each face.

a 

b 

c 

d 

e 

f 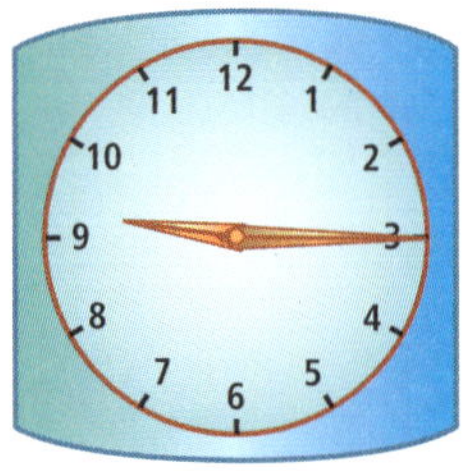

2 On each face show the time given.

a 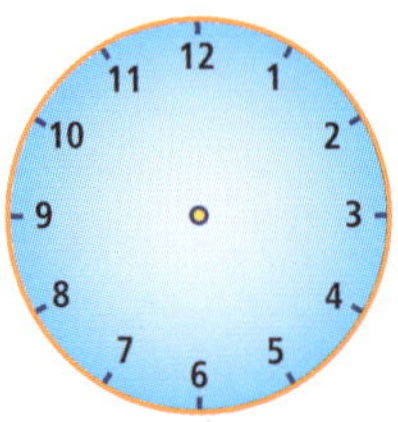

a quarter past 10

b 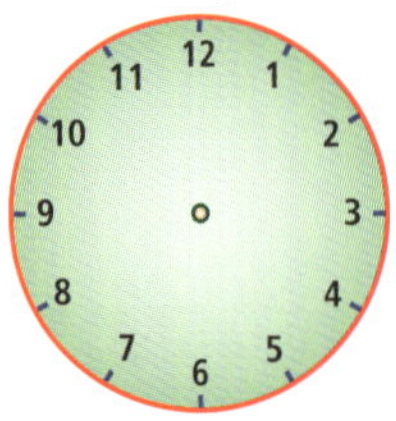

a quarter to 8

c 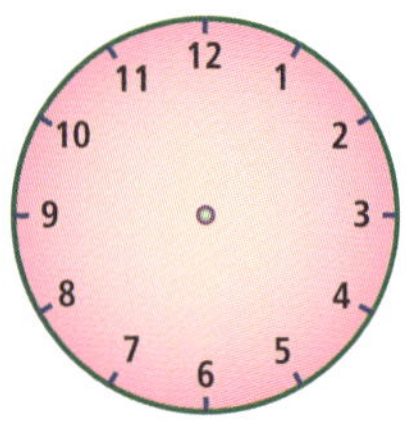

a quarter to 4

d 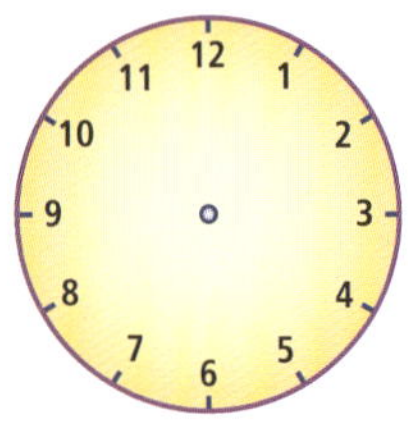

half past 7

e 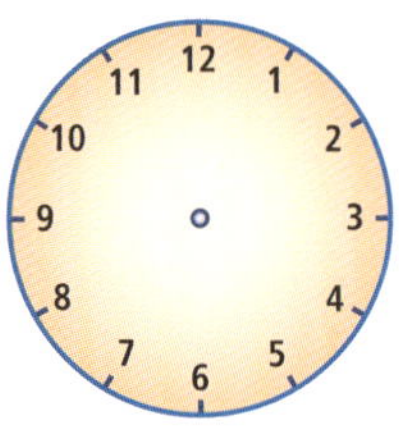

a quarter past 8

f 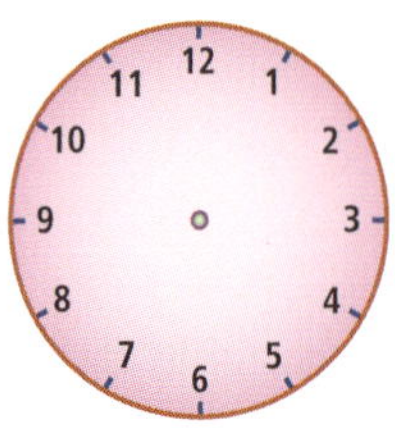

half past 2

g 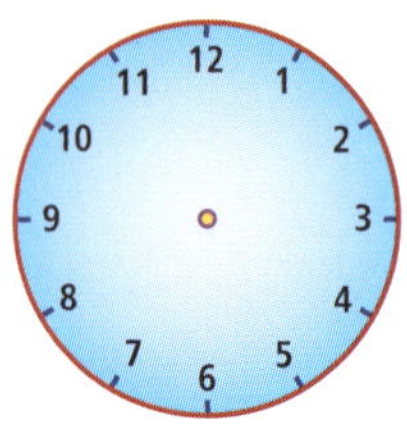

a quarter to 1

h 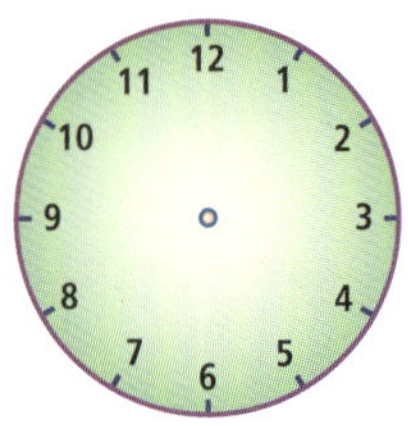

a quarter past 2

 • *AUSTRALIAN SIGNPOST MATHS 2* • ISBN 978 1 4886 2181 9

17D Months of the Year

The months of the year have a particular order.

January, February, March, April,
May, June, July, August, September,
October, November, December

1 Which month comes after:

a February? b August?

c October? d December?

2 How many months from the beginning of January to the beginning of:

a May? b October? c August?

d July? e March? f June?

g April? h December? i September?

3 Which month comes before:

a January? b June?

c September? d March?

4 Write the three months that begin with J.

5 Write the two months that begin with M.

6 Use the internet to find the months in which these events occur.

a Chinese New Year b Valentine's Day

c Mother's Day d NAIDOC Week

18A How Many More?

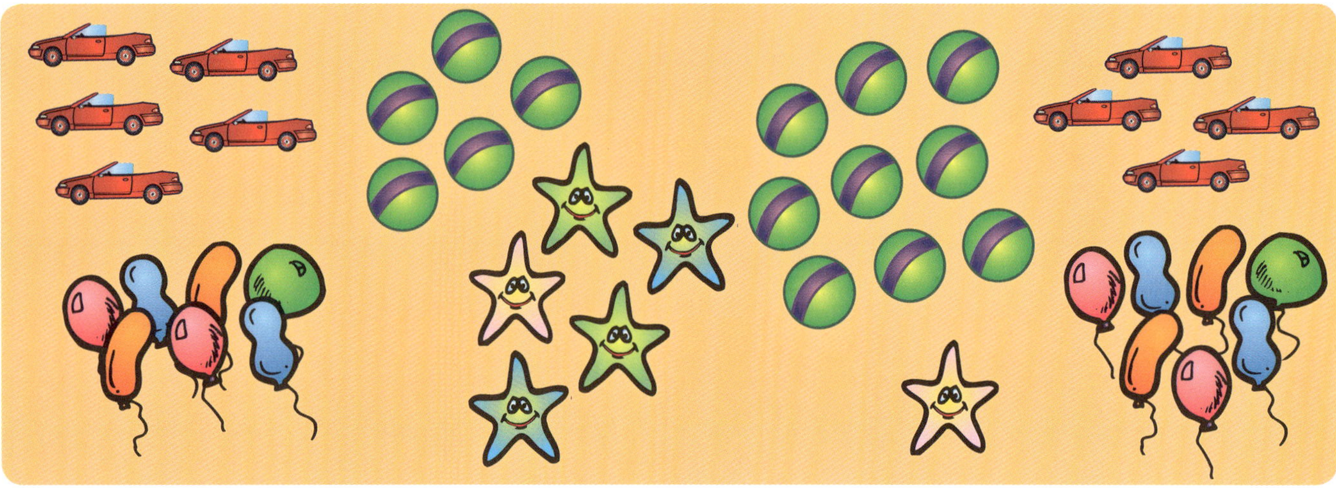

1 In this picture, how many more would I need to make 20:

a cars? 5 + 4 + ___ = 20

b balloons? ___ + ___ + ___ = 20

c balls? ___ + ___ + ___ = 20

d starfish? ___ + ___ + ___ = 20

2

a 6 + ___ = 12

b 3 + 8 + ___ = 16

c 11 + ___ = 15

d 4 + ___ + 5 = 13

e ___ + 9 = 13

f 10 + ___ + 5 = 19

g 5 + ___ = 12

h 8 + 9 + ___ = 20

i ___ + 8 = 16

j 7 + 3 + ___ = 20

3 List all possible answers.

a

6 + ___ + ___ = 10

6 + ___ + ___ = 10

6 + ___ + ___ = 10

6 + ___ + ___ = 10

6 + ___ + ___ = 10

b

17 + ___ + ___ = 20

17 + ___ + ___ = 20

17 + ___ + ___ = 20

17 + ___ + ___ = 20

4 Explore possible answers for 10 + ___ + ___ = 20.

18B Flipping a Shape

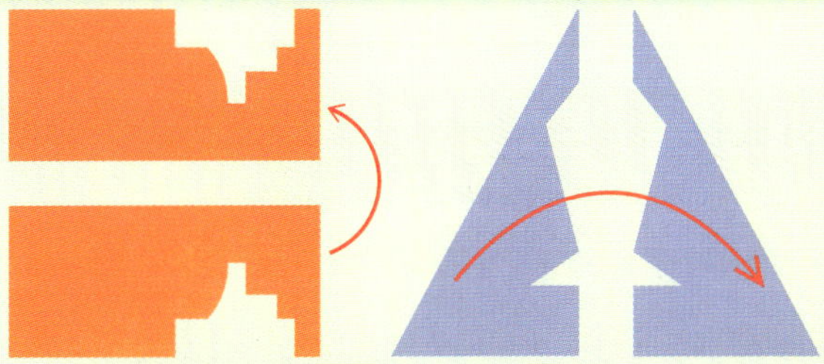

Flip We can flip a shape over in any direction.

1. Circle the pictures that show a flip.

2. Trace each shape. Flip the shape across the page to finish the pattern.

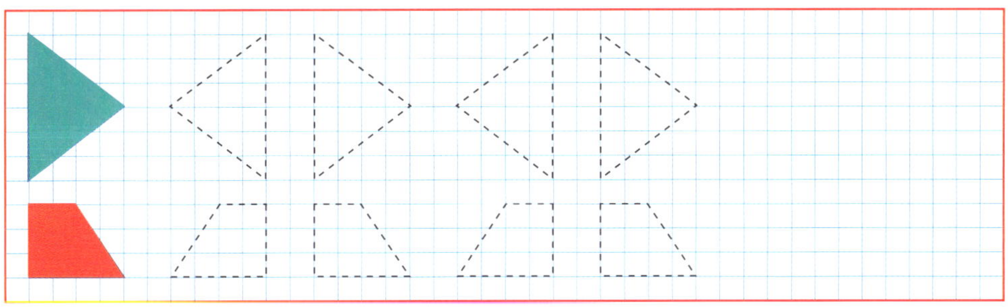

Lay two pattern blocks together.
Place two more blocks near them that would be a 'flip' of the first two blocks.
For example:

Slide back the edge of this page to see the edge of page 73 beside it.

 • *AUSTRALIAN SIGNPOST MATHS 2* • ISBN 978 1 4886 2181 9

Chance

1 Link each event with the better label.

a

more likely

less likely

b

more likely

less likely

c

more likely

less likely

2 Complete this sentence.
It is more likely that I will ______________________

____

3 Sam will draw out one counter. Match the boxes.

 or 

Certain means it must happen.

impossible | likely | certain | unlikely

 AUSTRALIAN SIGNPOST MATHS 2 • ISBN 978 1 4886 2181 9

18D Sliding a Shape

CONCEPT

Slide We can slide a shape in any direction.

1. Circle the pictures that show a slide.

2. Trace each shape. Slide the shape across the page to finish the pattern.

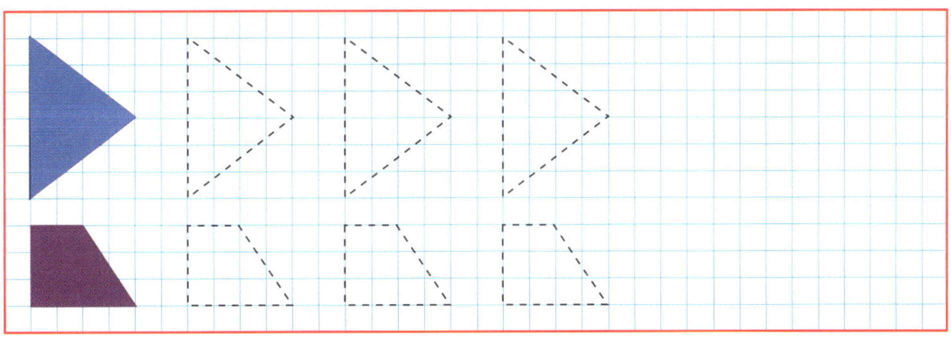

Look at these pictures beside the edge of page 71.

FUN SPOT

Play **Simon Says**, giving directions that include 'slide', 'forwards', 'backwards', 'left' and 'right'.

Slide 2 steps backwards.

 • • ISBN 978 1 4886 2181 9

19A Numbers to 1000

CONCEPT

❶ Complete each part.

a
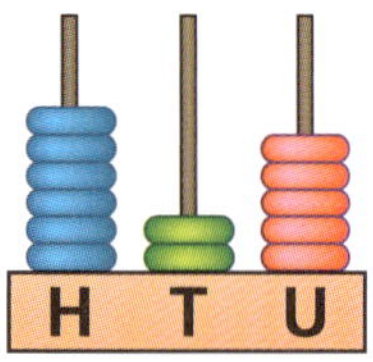

b
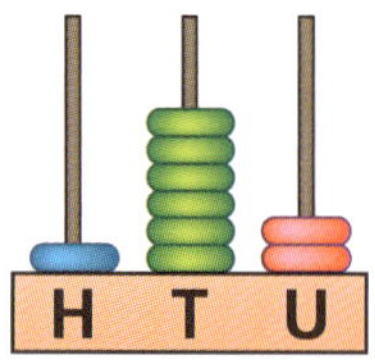

c
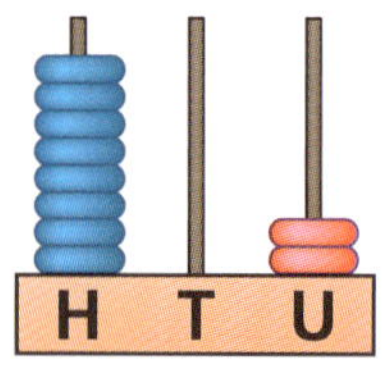

d
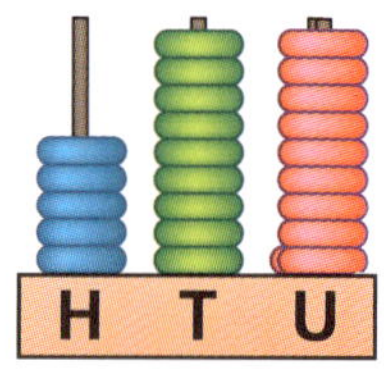

e
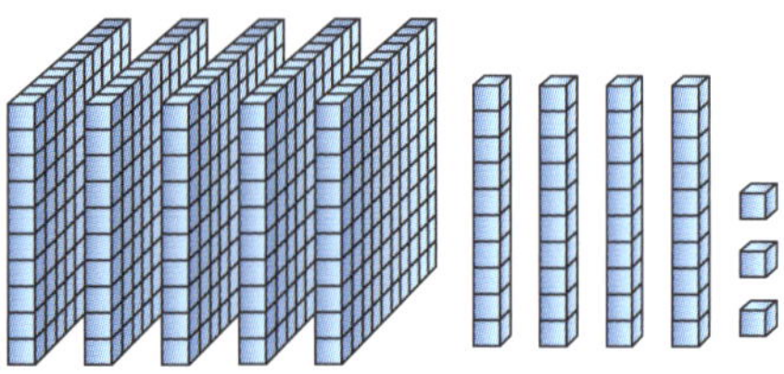

☐ hundreds + ☐ tens + ☐ ones

Hundreds	Tens	Ones
☐	☐	☐

= ☐

f
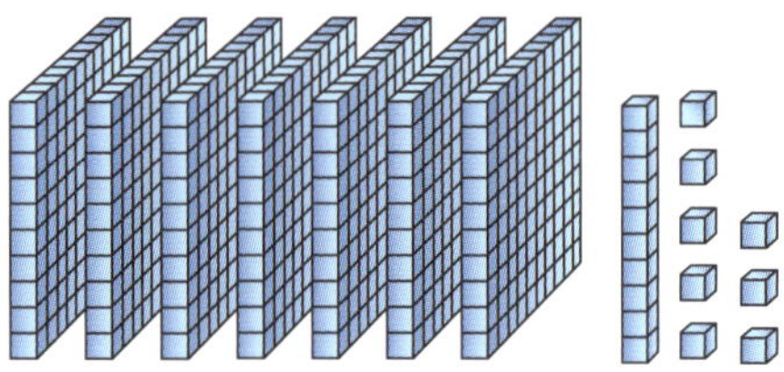

☐ hundreds + ☐ tens + ☐ ones

Hundreds	Tens	Ones
☐	☐	☐

= ☐

g
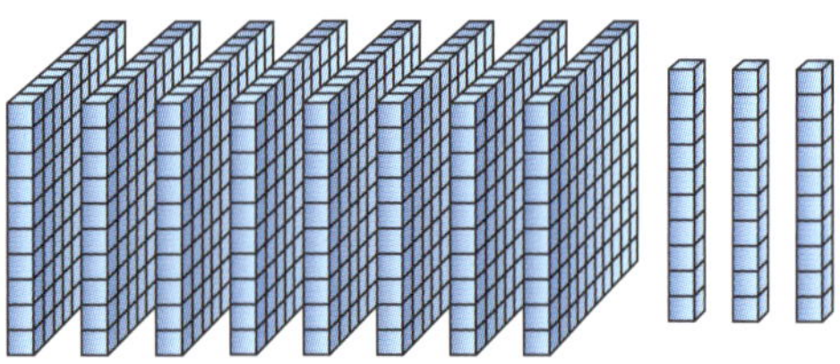

☐ hundreds + ☐ tens + ☐ ones

Hundreds	Tens	Ones
☐	☐	☐

= ☐

INVESTIGATION

Use place-value blocks to partition and regroup these numbers. Explain your answer to a friend. For example, 500, 5 hundreds, 50 tens, 500 ones, 100 fives.

- 800
- 300
- 700
- 400

 • *AUSTRALIAN SIGNPOST MATHS 2* • ISBN 978 1 4886 2181 9

19B Number Patterns

1 Complete the number patterns and describe them.

a 5, 10, 15, ___, ___, ___, ___, ___

b 10, 20, 30, ___, ___, ___, ___, ___

c 20, 18, 16, ___, ___, ___, ___, ___

2 On the hundred chart:

a count by twos to fifty.

b count by fives to fifty and circle each number counted.

c count by tens to one hundred and colour each number counted.

d count by fours to forty.

Discuss the patterns. Discuss how the fives and tens patterns are similar.

1	2	3	4	5	6	7	8	9	10
11	12	13	14	15	16	17	18	19	20
21	22	23	24	25	26	27	28	29	30
31	32	33	34	35	36	37	38	39	40
41	42	43	44	45	46	47	48	49	50
51	52	53	54	55	56	57	58	59	60
61	62	63	64	65	66	67	68	69	70
71	72	73	74	75	76	77	78	79	80
81	82	83	84	85	86	87	88	89	90
91	92	93	94	95	96	97	98	99	100

3 Write the missing numbers. Describe the rule for each.

a 4, 9, 14, ___, 24, ___

b 70, 60, 50, ___, 30, ___

19C Balance Scales

CONCEPT

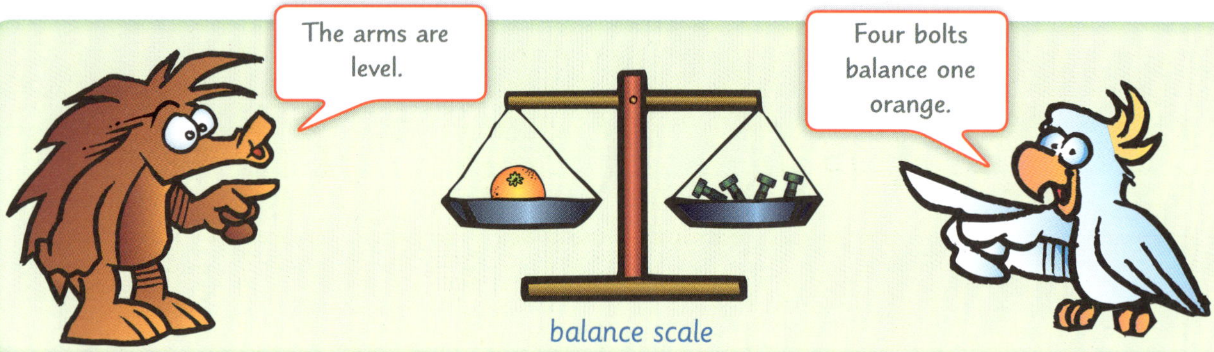

ACTIVITY

1 Use a balance scale to find answers to these.

a 2 marbles balance ☐ craft sticks.

b 1 ball balances ☐ marbles.

c 8 pens balance ☐ marbles.

d 4 pens balance ☐ craft sticks.

e 1 tennis ball balances ☐ blocks.

f 5 marbles balance ☐ blocks.

2 a Place 8 screws in a balance pan. What will balance the screws? Draw what you found.

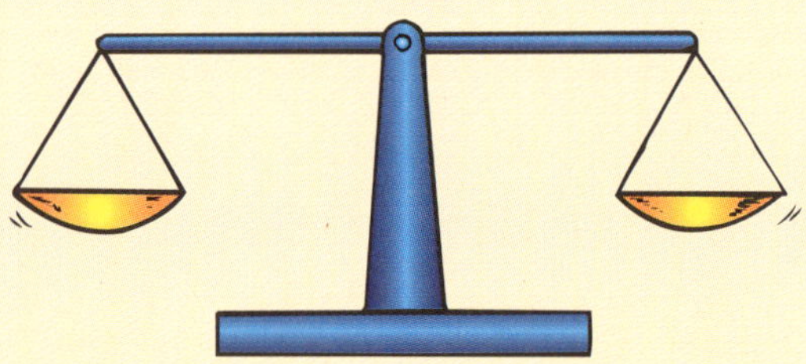

b Place a glue stick in a balance pan. What will balance the glue stick? Draw what you found.

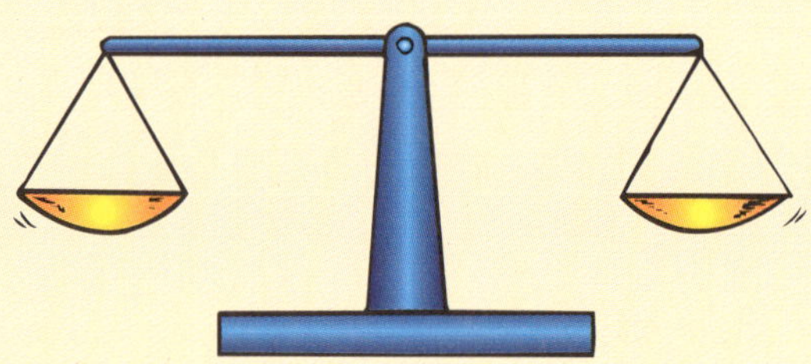

3 Write a balance story about your findings.

a

b

© PEARSON AUSTRALIA 2018 • *AUSTRALIAN SIGNPOST MATHS 2* • ISBN 978 1 4886 2181 9

19D Ordering Masses

1 a Handle three objects like these and estimate which one is the lightest and which one is the heaviest.

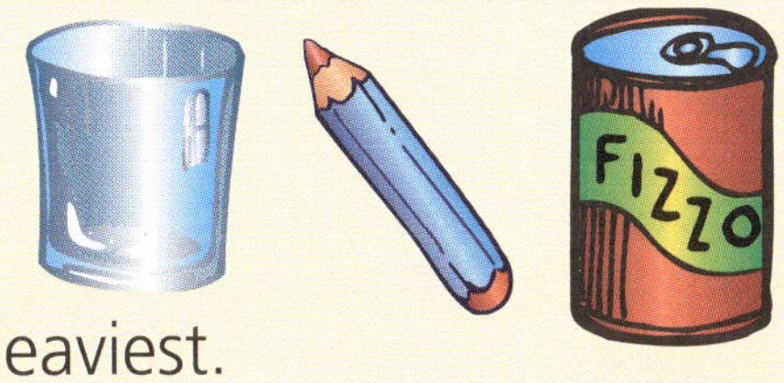

b Order the objects from lightest to heaviest.

_______ (lightest), _______ , _______ (heaviest)

c Use blocks and a balance scale to find the mass of each object.

Object	Mass
	blocks
	blocks
	blocks

d Write a sentence about the masses of the three objects.

____

____

2 Choose other sets of three objects and compare them in the same way. Record your results.

Object	Mass
	marbles
	marbles
	marbles

Object	Mass

Object	Mass

Object	Mass

20A Value of Coins

❶ Colour the coin you would use to pay for each item.

a

b

c

d

e

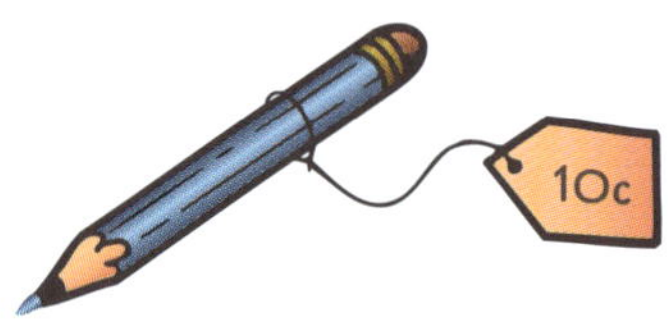

❷ Write the **value** of each coin in order.

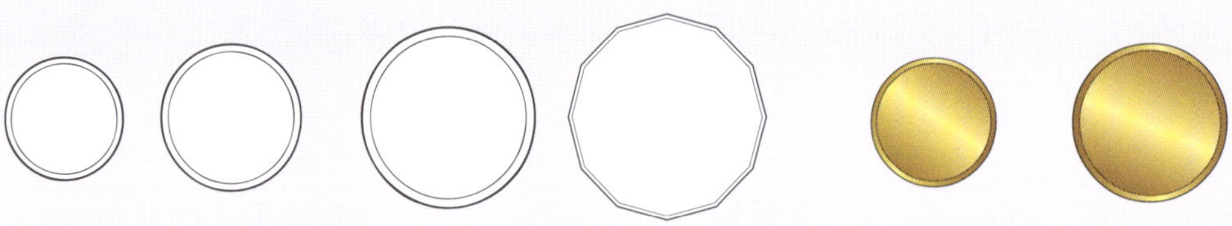

 • *AUSTRALIAN SIGNPOST MATHS 2* • ISBN 978 1 4886 2181 9

20B Value of Coins

1 Colour the coins you would use to pay for each item.

a

b

c

d

2 Use skip counting to find each total.

a  ☐ cents

b  ☐ cents

3 Write the value of each coin.

a  ☐

b  ☐

c  ☐

 • *AUSTRALIAN SIGNPOST MATHS 2* • ISBN 978 1 4886 2181 9

20C Comparison of Areas

CONCEPT

Area is the measure of the amount of surface.

The red square has more area than the green square.

1. Which coloured shape has the larger area?

a 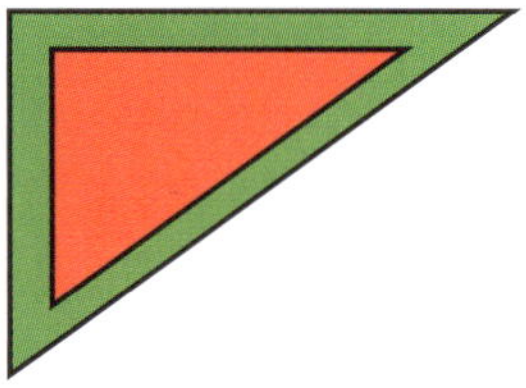

b 

c 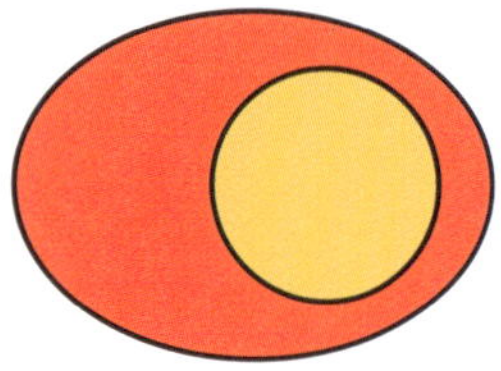

d 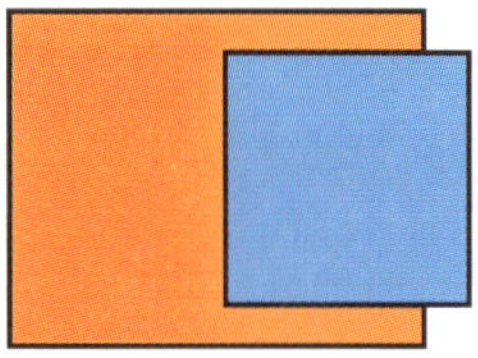

e 

f 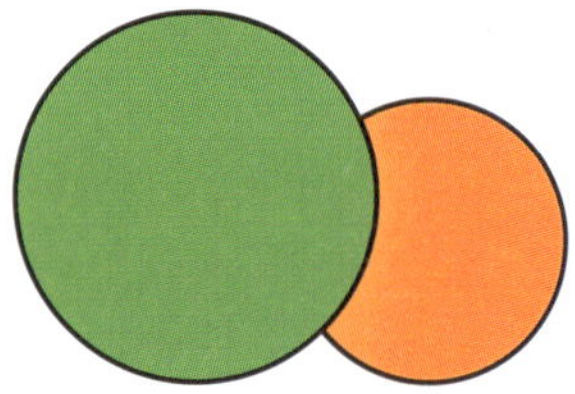

g 

h 

i 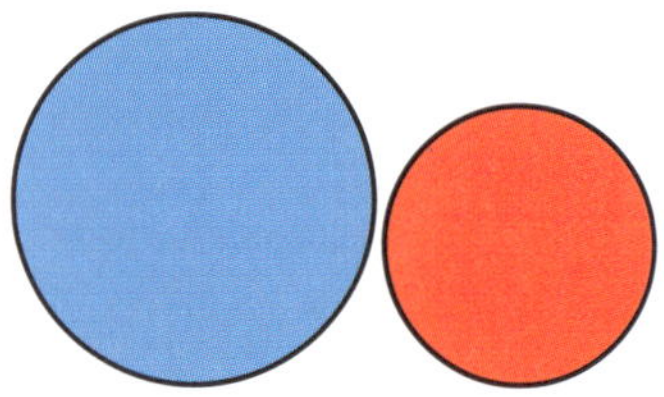

INVESTIGATION

- Choose four flat objects.

  By laying one object on top of another, order their areas from smallest to largest.

20D Area Using Informal Units

The area of shapes or objects can be compared using uniform units.

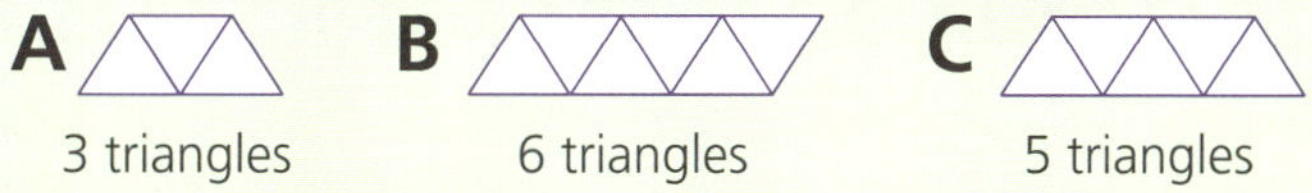

B has the greatest area.

CONCEPT

1 Use the square informal unit to find the area of each shape.

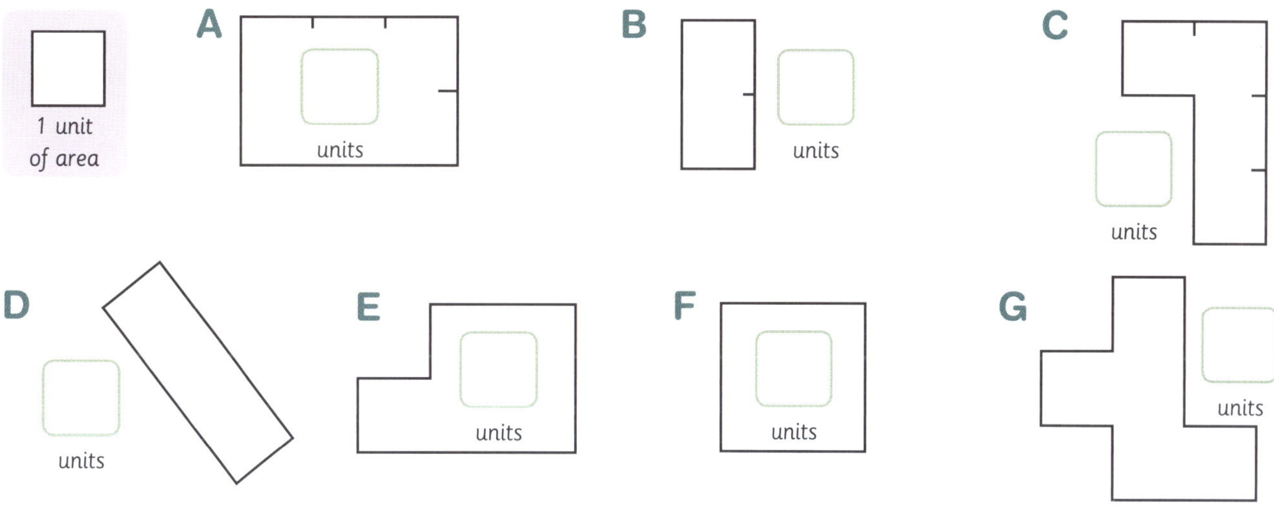

2 Draw a shape with an area of 8 square units.

a Which shape has the smallest area?

b Which shape has the largest area?

c Which shape has the same area as F?

d Which shape has the same area as E?

e Which shape has double the area of shape D?

ACTIVITY

Use place-value blocks as a uniform unit to compare three different areas.

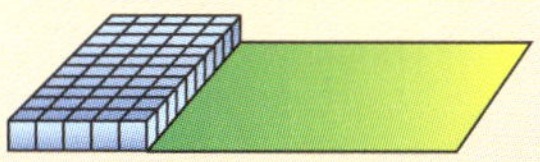

21A Multiplication Sign

1 Match each group with the correct number sentence.

× means 'groups of'.

- 2 × 9 = 18
- 5 × 2 = 10
- 6 × 3 = 18
- 2 × 6 = 12
- 3 × 4 = 12
- 5 × 3 = 15

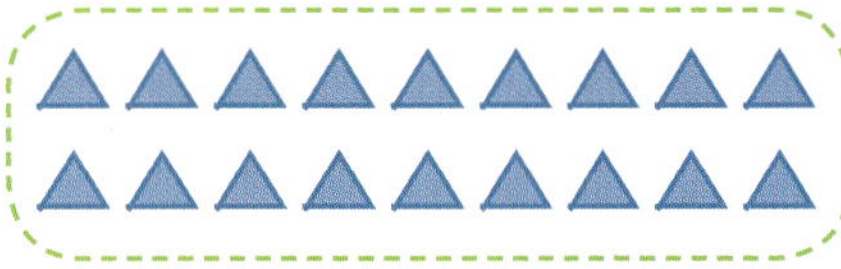

2 Complete these number sentences.

a		0 groups of 2 = ☐	**b**	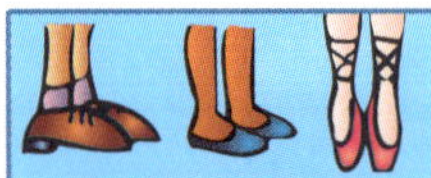	☐ × ☐ = ☐
c		1 group of 2 = ☐	**d**		4 × 2 = ☐
e		2 groups of 2 = ☐	**f**		5 × 2 = ☐

3 Draw pictures to show each story. Complete the number sentences.

a 3 vases of 4 flowers each

☐ groups of ☐ = ☐

☐ × ☐ = ☐

b 2 boxes with 6 balls in each

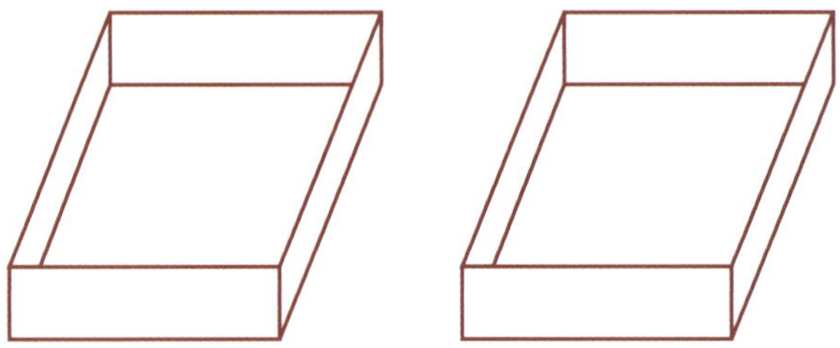

☐ groups of ☐ = ☐

☐ × ☐ = ☐

4 Use counters to make arrays to answer these questions.

a 3 × 7 = ☐ **b** 4 × 6 = ☐ **c** 5 × 3 = ☐

 • *AUSTRALIAN SIGNPOST MATHS 2* • ISBN 978 1 4886 2181 9

21B Multiplication Sign

1 Complete the number sentences.

× means 'groups of' or 'rows of'.

a

___ × ___ = ___

b

___ × ___ = ___

c

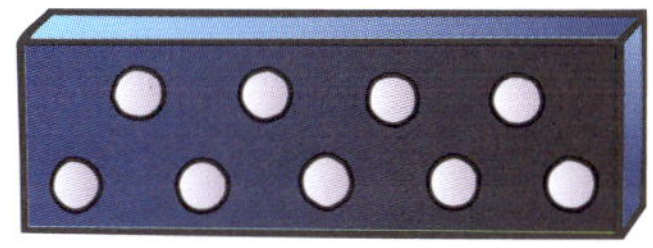

___ × ___ = ___

d

___ × ___ = ___

e

___ × ___ = ___

f

___ × ___ = ___

g

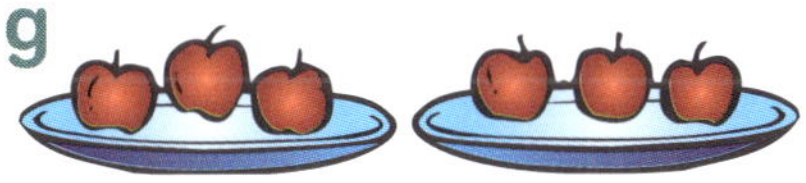

___ × ___ = ___

h

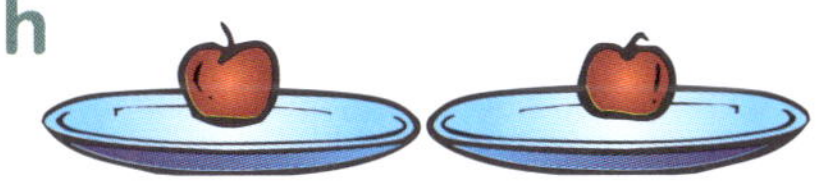

___ × ___ = ___

i

___ × ___ = ___

2 Draw pictures to show these.

a

b

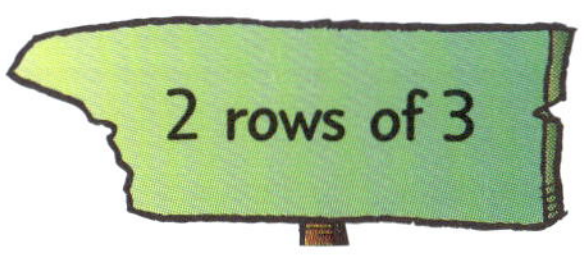

___ × ___ = ___

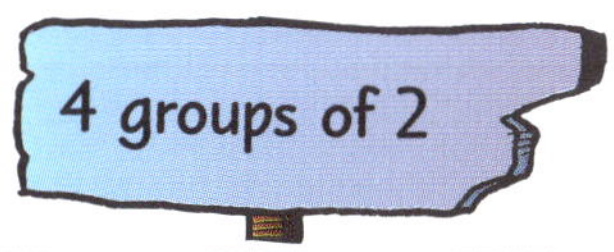

___ × ___ = ___

3 Use the dots to help you complete the number sentences.

2 × 6 = ___

3 × 4 = ___

4 × 3 = ___

6 × 2 = ___

21C The Parallelogram

1 Name the shapes used.

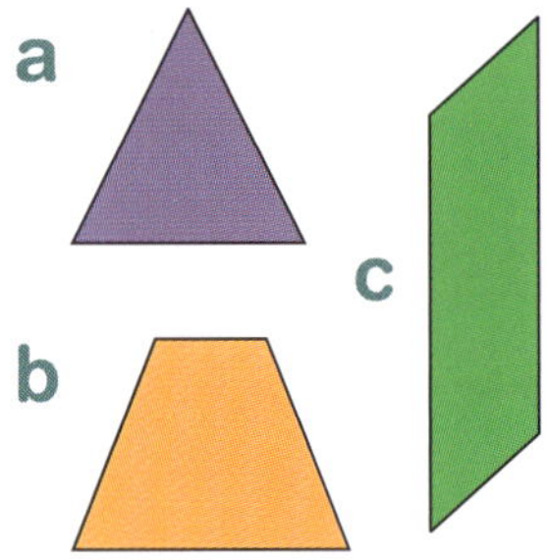

a

b

c

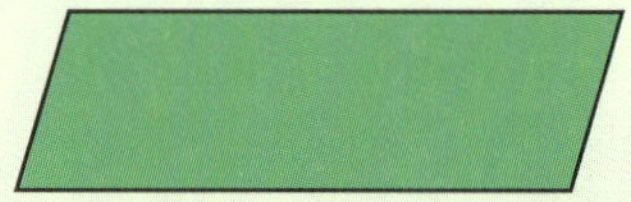

CONCEPT

This is a **parallelogram**. It has two pairs of parallel sides.

circle rhombus triangle square rectangle hexagon trapezium

2 Trace each shape. Write its name.

3 Colour the parallelograms red.

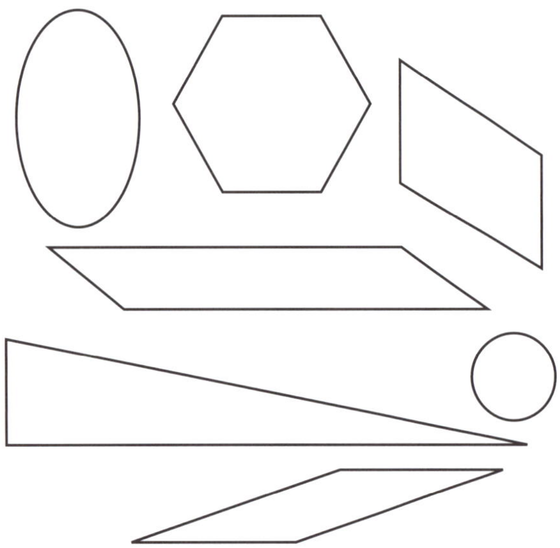

4 Draw two parallelograms.

5 a How many sides has a parallelogram?

b How many corners has a parallelogram?

c Does a parallelogram have 2 pairs of parallel sides?

 • *AUSTRALIAN SIGNPOST MATHS 2* • ISBN 978 1 4886 2181 9

21D Using Tally Marks

We can count using tally marks.

卌 II = 7

This means 5.

1 Write the number shown by each tally.

a ||| ☐ b 卌 | ☐ c 卌 卌 ☐

d 卌 |||| ☐ e 卌 卌 | ☐ f 卌 | ☐

2 Cross off one ball at a time and put a tally mark in the table.

Christmas Balls	
red	
blue	
green	
orange	

How many of the balls are:

a red? ☐ b blue? ☐ c green? ☐

d orange? ☐ e How many balls altogether? ☐

Make a Tally

Complete this tally for the people in your class. You can decide who has light hair and who has dark hair.

Hair Colour	
light	
dark	

22A Building to the Next 10

17 + 5 = 17 + 3 + 2 = 20 + 2 = 22

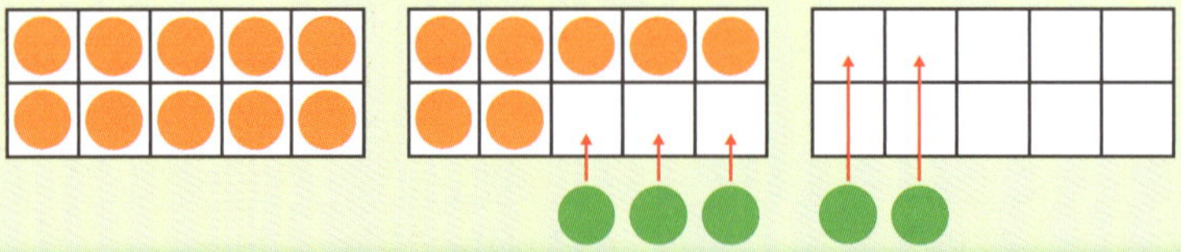

1. Use the ten frames to build to the next 10.

a 18 + 4 = 18 + ☐ + ☐ = 20 + ☐ = ☐

b 17 + 6 = 17 + ☐ + ☐ = 20 + ☐ = ☐

c 19 + 5 = 19 + ☐ + ☐ = 20 + ☐ = ☐

Use ten frames to make up questions of your own.

2. Help Banjo find the answers using the 'build to the next 10' strategy.

19 + 6 =

19 + 4 =

19 + 3 =

16 + 8 =

16 + 6 =

17 + 4 =

17 + 7 =

18 + 5 =

18 + 3 =

18 + 6 =

18 + 7 =

 • AUSTRALIAN SIGNPOST MATHS 2 • ISBN 978 1 4886 2181 9

22B Building to the Next 10

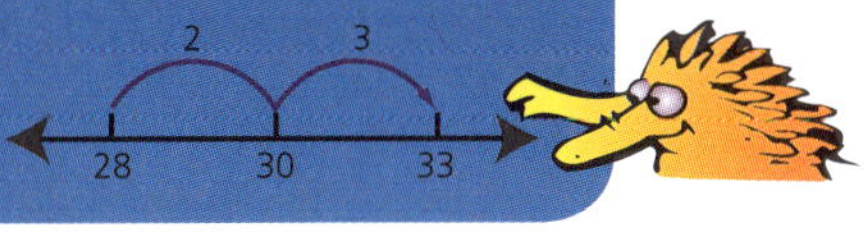

CONCEPT

$28 + 5 = 28 + 2 + 3$

$= 30 + 3$

$= 33$

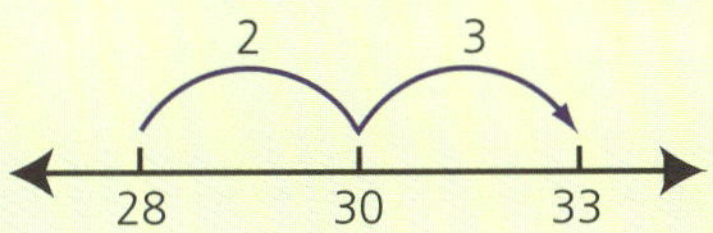

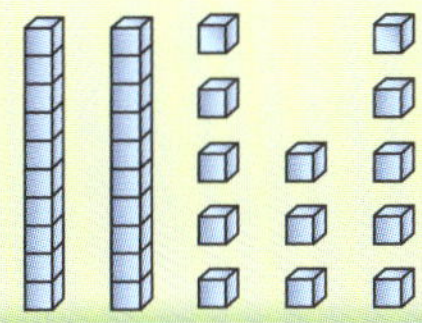

1 **a** 37 + 6 = 37 + 3 + ___
= 40 + ___
= ___

b 25 + 7 = 25 + 5 + ___
= 30 + ___
= ___

c 44 + 8 = 44 + 6 + ___
= 50 + ___
= ___

d 73 + 9 = 73 + 7 + ___
= 80 + ___
= ___

2 **a** 56 + 5 = 56 + 4 + ___
= ___ + ___
= ___

b 69 + 4 = 69 + 1 + ___
= ___ + ___
= ___

c 38 + 6 = 38 + 2 + ___
= ___ + ___
= ___

d 84 + 8 = 84 + 6 + ___
= ___ + ___
= ___

3 Try to do these in your head.

a 38 + 3 = ___ **b** 54 + 7 = ___

c 77 + 8 = ___ **d** 29 + 5 = ___

e 46 + 9 = ___ **f** 55 + 7 = ___

g 35 + 5 = ___ **h** 89 + 4 = ___

Ask: What is needed to reach the next 10?

 • *AUSTRALIAN SIGNPOST MATHS 2* • ISBN 978 1 4886 2181 9

22C Pyramids and Prisms

CONCEPT

Pyramid

One flat base, all other sides are triangles.

Prism

Both ends are the same, other sides are rectangles.

1. Write **prism**, **pyramid** or **cylinder** under each object.

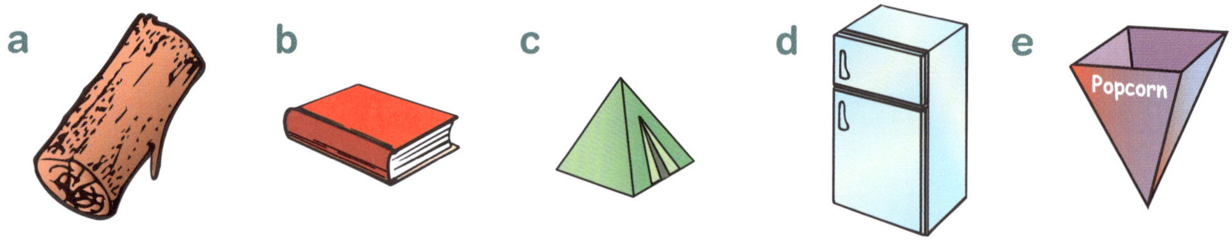

2. Use words like **smooth**, **corner**, **edge**, **face** (flat surface) and **curved surface** to describe each object.

a

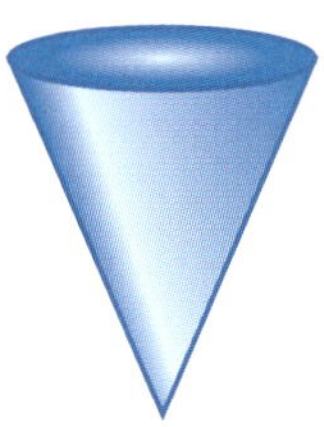

b

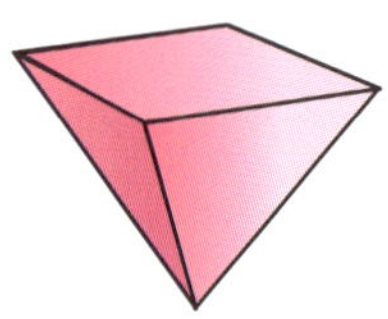

3. Complete the pattern of pyramids and prisms.

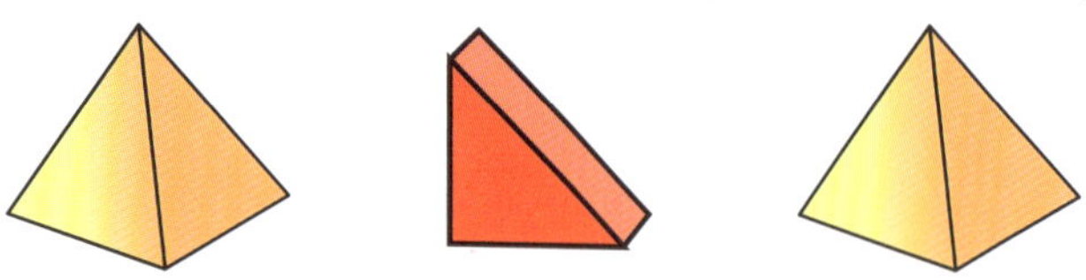

ACTIVITY

Use Plasticine or playdough to make models of pyramids, prisms, cones and cylinders.

 • *AUSTRALIAN SIGNPOST MATHS 2* • ISBN 978 1 4886 2181 9

22D Recognising 3D Objects

1 Match each item with one of the 3D objects. Write the name of each 3D object below it.

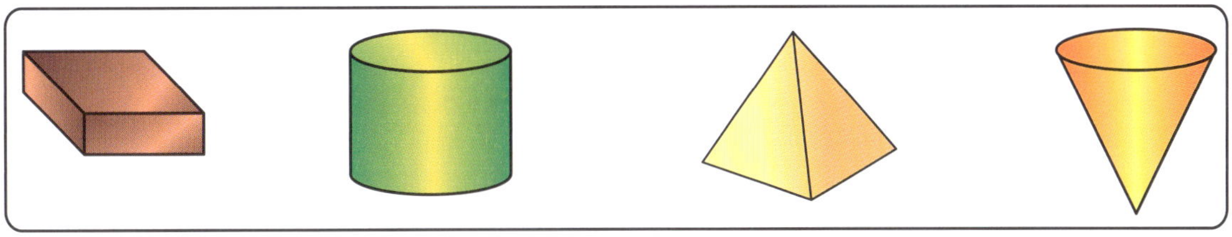

2 Draw a picture of the 3D object. How many surfaces does it have?

cylinder	☐ surfaces	cone	☐ surfaces

3 Complete the pattern of pyramids and cones.

Challenge

- Build a model of a cube using craft sticks or toothpicks. Use Blu Tack or jelly lollies to hold parts together.
- Build a model of a real object using 3D objects.

23A Australian Coins

CONCEPT

The order of size is not the order of value.

1 Write the values of the coins in the order above.

2 **a** James has ____.

b Sara has ____.

c ____ has less.

d ____ has more.

3 How much money?

a

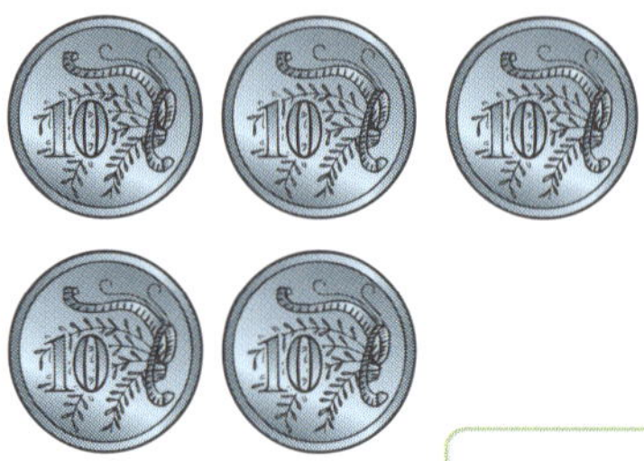

____ cents

b

____ cents

c

____ cents

d

____ cents

23B Money

To find 'how much', start with the highest value and count on.

$1 … $1.50 … $1.70 … $1.80

1 How much money?

a

b

c

d

2 Colour the coins you would use to pay for each item.

a

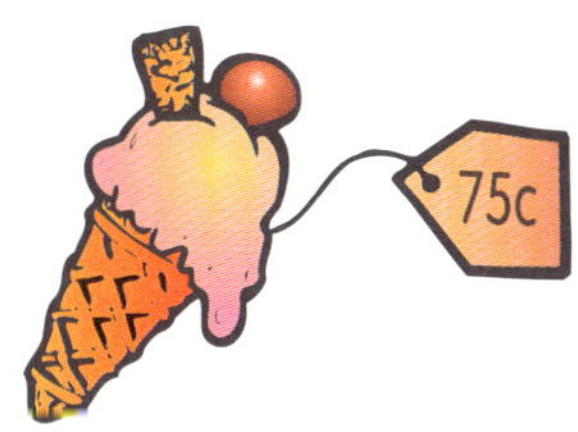

b

 • *AUSTRALIAN SIGNPOST MATHS 2* • ISBN 978 1 4886 2181 9

23C Making Graphs

Look at this picture.
Without counting, estimate which group is largest.

1. In these squares draw one wombat to represent each wombat in the picture. (Cross out each wombat in the picture after you redraw it.) Do the same for the kangaroos and birds. →

Animals Seen		
Birds	Kangaroos	Wombats

2. Use the drawings in the squares to compare the groups.

   a Which group is the smallest? ☐

   b Which group is the largest? ☐

   c How many more wombats are there than birds? ☐

   d What is the difference between the number of kangaroos and the number of birds? ☐

3. Complete this table to record the animals shown in the picture above.

Animals Seen	Number
Birds	
Kangaroos	
Wombats	

- Draw a table showing the people in your classroom.

Boys	
Girls	
Adults	

23D Tally Marks

1 Nina used toothpicks to find the number of children in 30 families.

Number of Children in Families	
1 child	| | | | | |
2 children	| | | | | | | | | | | | |
3 children	| | | | |
4 children	| | | |
5 children	| |

Number of Children in Families	
1 child	
2 children	
3 children	
4 children	
5 children	

a How many families had 5 children?

b How many families had 1 child?

c What was the most popular number?

d How many families had 6 children?

CONCEPT

||| → 3
|||| → 4
𝍸 → 5
𝍸 | → 6

2 Julie recorded the number of vowels used in her story.

a Which vowel was used the most?

b Which vowel was used the least?

c How many times was the letter **e** used?

d How many more times was **e** used than **a**?

e How many vowels were used altogether?

f Is it easier to count separate strokes or groups of 5?

Vowels Used

a	𝍸 𝍸 𝍸 𝍸
e	𝍸 𝍸 𝍸 𝍸 𝍸
i	𝍸 𝍸 ||||
o	𝍸 𝍸 |||
u	𝍸 |

INVESTIGATION

Take a handful of coloured counters and use them to complete this tally.

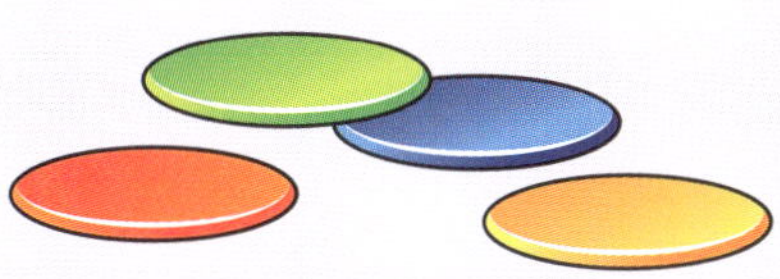

Colour	Tally	Total

24A Number Patterns

1 Start at 2. Colour every 2nd number green. These are even numbers.

1	2	3	4	5	6	7	8	9	10
11	12	13	14	15	16	17	18	19	20
21	22	23	24	25	26	27	28	29	30

2 Describe the pattern shown on each number line.

a

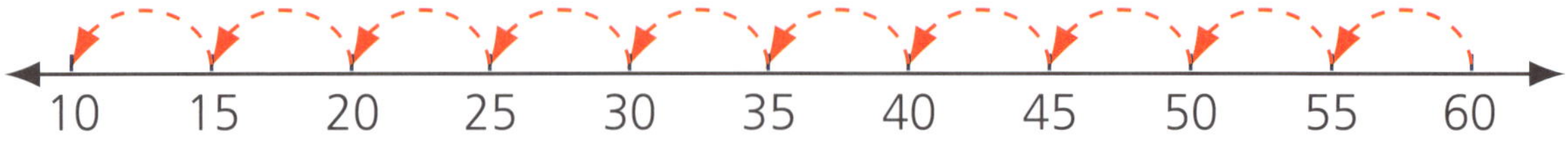

b

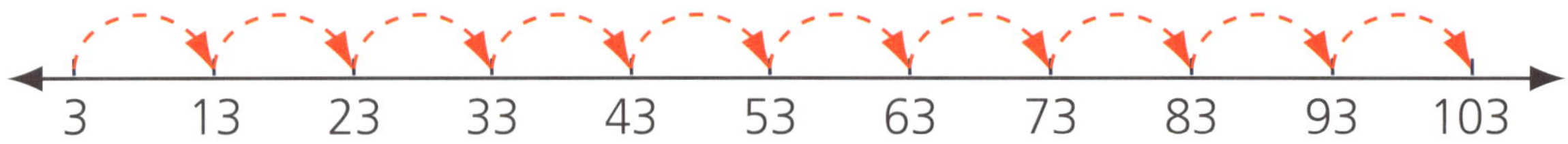

3 Write the missing numbers. Describe the rule for each pattern.

a

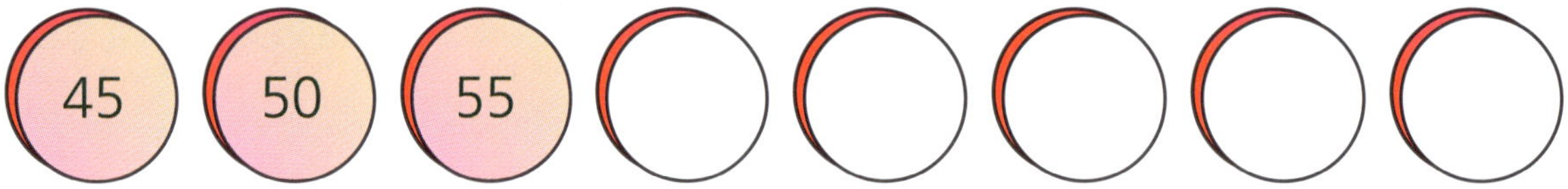

b

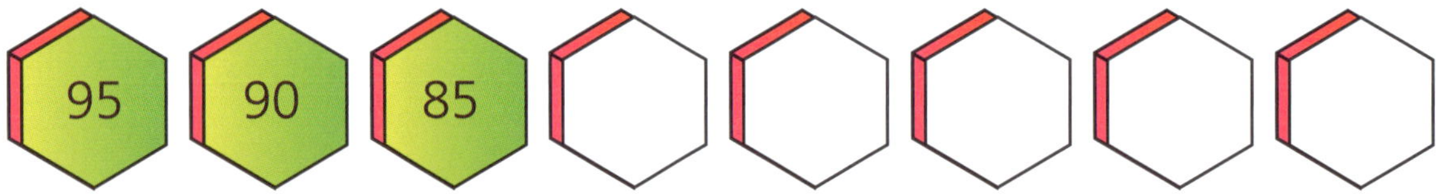

c

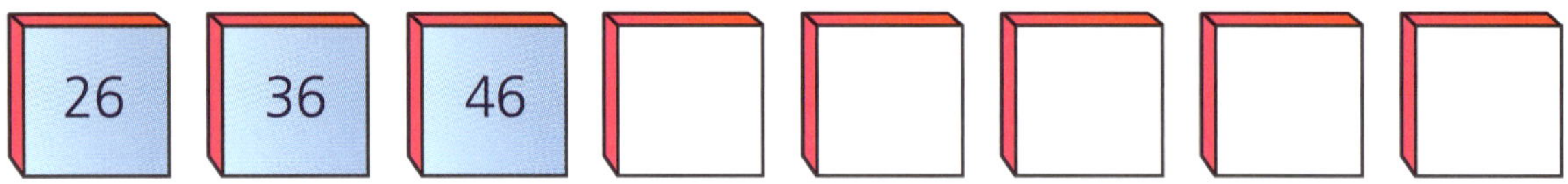

4 Write in the next three numbers in each number pattern.

a 64, 54, 44,

b 65, 70, 75,

c 90, 80, 70,

d 37, 47, 57,

e 15, 25, 35,

f 34, 36, 38,

Investigate the effect of repeatedly adding or subtracting 10.

 ISBN 978 1 4886 2181 9

24B Amounts to $2

1 How much money is in each moneybox?

a

b

CONCEPT

$1 + 20c + 5c = $1.25

To find ‘how much’, start with the highest coin and count on.

c

d

e

2 What is the value of each set of coins?

a

b

c

d

INVESTIGATION

- Use coins or play money to make 50c in as many ways as you can. My number of ways = ____.
- Use coins to make amounts chosen by your partner.

 • *AUSTRALIAN SIGNPOST MATHS 2* • ISBN 978 1 4886 2181 9

Fractions

1 Complete this table. Write 'less than half' or 'more than half' for each.

	Object	Number Coloured	Total in Group	Statement (part coloured)
a		3	7	less than half
b				
c				
d				
e				

2 Colour part of each shape to match the fraction.

a 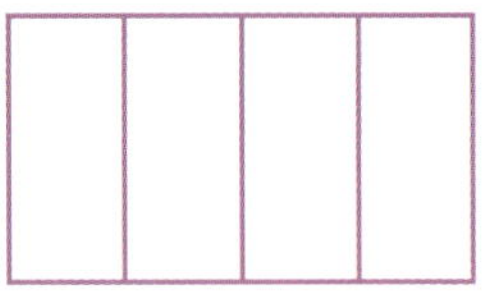
$\frac{1}{4}$

b 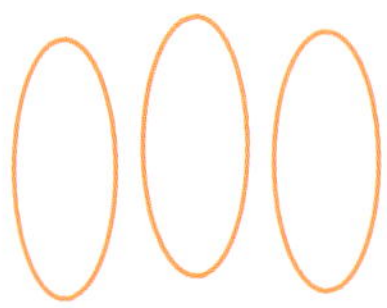
$\frac{1}{3}$

c 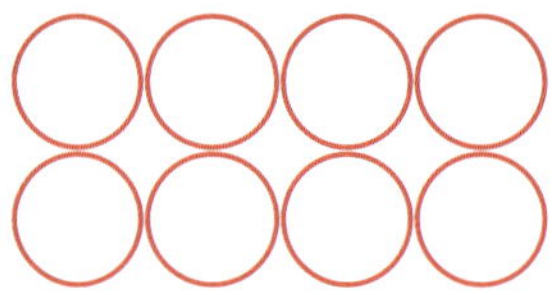
$\frac{1}{8}$

d 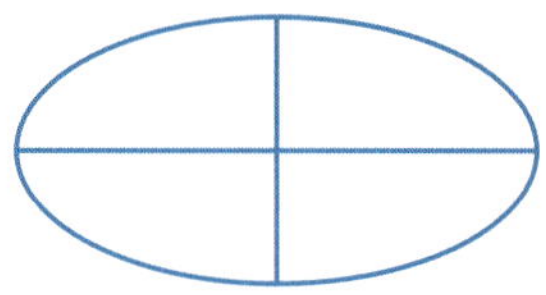
$\frac{1}{2}$

3 Circle the objects that are **not** divided into halves.

4 Without counting, circle half of each group.

Count to check.

 ISBN 978 1 4886 2181 9

24D Giving Directions

❶

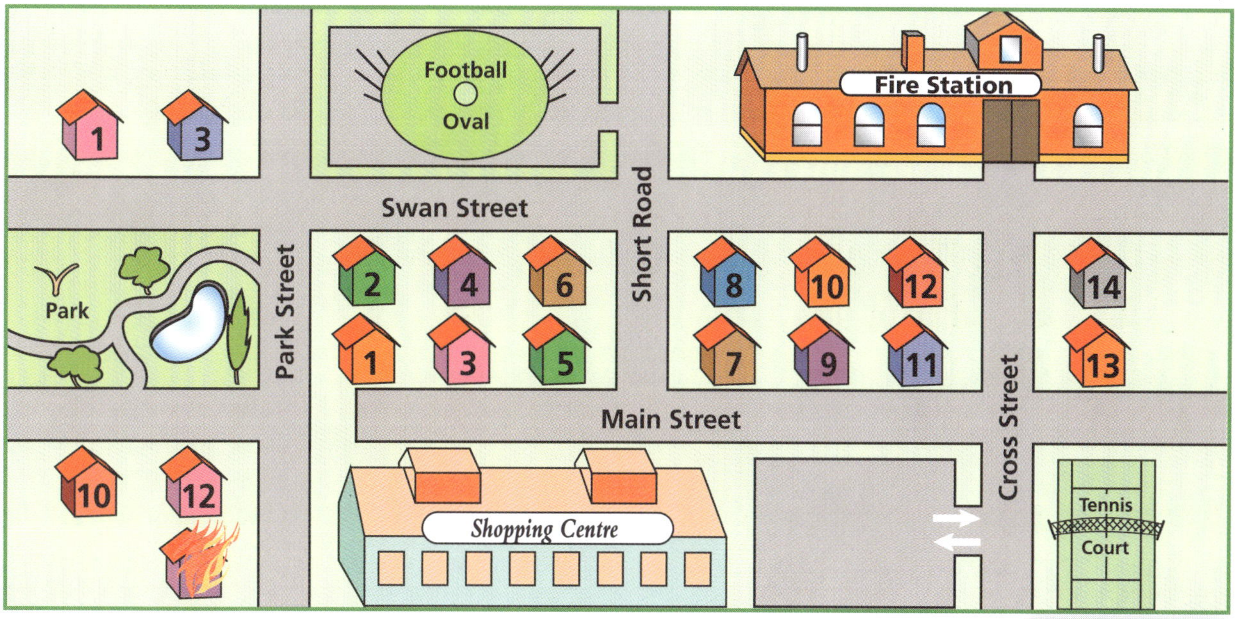

a Trace the shortest route from the fire station to the burning house. Write instructions.

Discuss other paths.

____

____

b Write instructions for getting from 11 Main Street to the football oval.

____

____

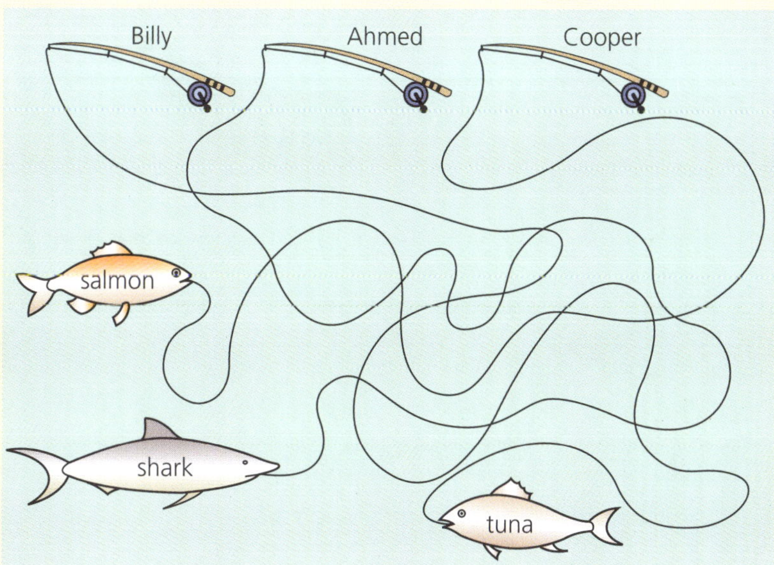

FUN SPOT

❷ Follow each line until you come to a fish.

What did they catch?

Billy ____________

Ahmed ____________

Cooper ____________

25A Multiplication Problems

1 Use the drawings above to complete these number sentences.

a ☐ groups of ☐ flying birds = ☐ flying birds

b ☐ group of ☐ cows = ☐ cows

c ☐ × ☐ frogs = ☐ frogs

d ☐ × ☐ cockatoos = ☐ cockatoos

e 1 × ☐ fish = ☐ fish

f ☐ × ☐ turtles = ☐ turtles

g ☐ × ☐ trees = ☐ trees

2 Use the groups above to make up other problems.

a 4 groups of ☐ cockatoos = ☐ cockatoos

b ☐ groups of ☐ ☐ = ☐ ☐

c ☐ groups of ☐ ☐ = ☐ ☐

25B Multiplication

3 × 3 = 9

This is called a number sentence.

1 Write a number sentence for each part.

a

___ × ___ = ___

b

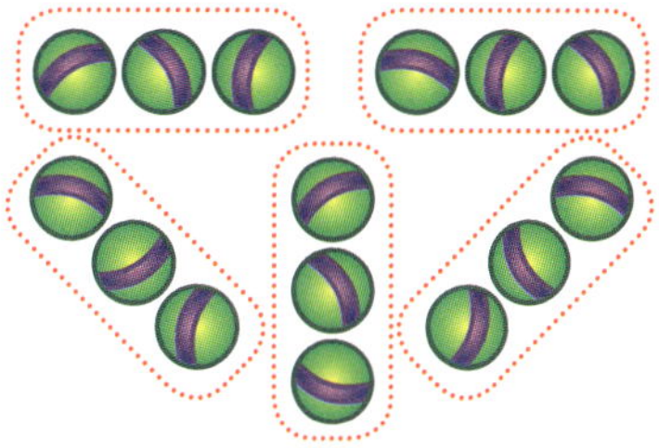

___ × ___ = ___

c

___ × ___ = ___

d

___ × ___ = ___

2 Write four different number sentences about these dogs.

a ___ × 2 = ___

b ___ × 4 = ___

c ___ × 8 = ___

d ___ × 16 = ___

3 Complete:

a Mina and Sam each ate 6 ice creams.

How many did they eat altogether? ___

___ × ___ = ___

b Four plates each held five biscuits.

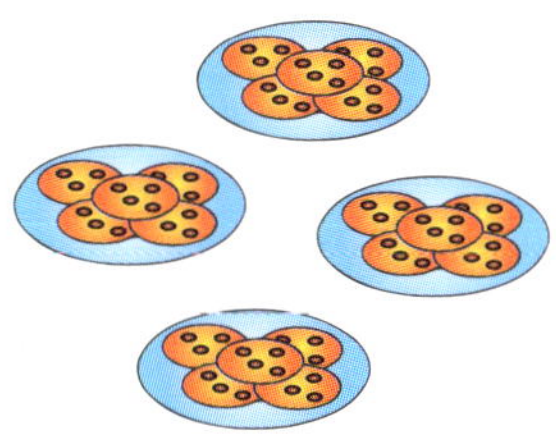

How many biscuits altogether? ___

___ × ___ = ___

 • *AUSTRALIAN SIGNPOST MATHS 2* • ISBN 978 1 4886 2181 9

25C Rounding

CONCEPT

Rounding off to the nearest ten.

1 a Is 12 closer to 10 or 20? ☐ b Is 77 closer to 70 or 80? ☐

c Is 21 closer to 20 or 30? ☐ d Is 89 closer to 80 or 90? ☐

e Is 33 closer to 30 or 40? ☐ f Is 106 closer to 100 or 110? ☐

2 Round to the nearest 10.

a 24 ☐	b 36 ☐	c 63 ☐	d 47 ☐
e 92 ☐	f 74 ☐	g 59 ☐	h 81 ☐
i 103 ☐	j 115 ☐	k 123 ☐	l 127 ☐

3 a Circle the numbers that round to 60.

57 58 67 68 64

61 69 62 65 52 56

b Circle the numbers that round to 90.

99 87 86 97 94 83

93 85 89 88 91

4 Write true (T) or false (F) for each statement.

a 124 is less than 130. ☐ b 127 is more than 117. ☐

c 111 is more than 101. ☐ d 60 is less than 47. ☐

e 35 is more than 76. ☐ f 93 is less than 100. ☐

25D Telling the Story from Data

CONCEPT

Every set of data tells a story.

List

Girl, Girl, Boy, Girl, Boy, Girl, Girl

Table

Boys	Girls
2	5

Graph: My Friends

Boy	Boy			
Girl	Girl	Girl	Girl	Girl

- Draw the graph.
- Draw the table.
- Tell the story.

Complete the graph and table, then tell the story.

1 Children at the party.

List

Boy, Girl, Boy, Boy, Girl, Boy, Boy, Girl

Graph:

At the Party

Boys	Girls

Table:

Boys	
Girls	

Tell the story.

____

____

____

2 Buttons on shirts.

List

3, 1, 4, 4, 2, 3, 3, 2, 4, 2, 3, 2, 1, 2, 5

Numbers of Buttons

1					
2					
3					
4					
5					

1 button	2 buttons	3 buttons	4 buttons	5 buttons

Tell the story.

____

____

26A Split Strategy (Addition)

CONCEPT

32 + 45

Add the tens, add the ones.

= (30 + 40) + (2 + 5)

= 77

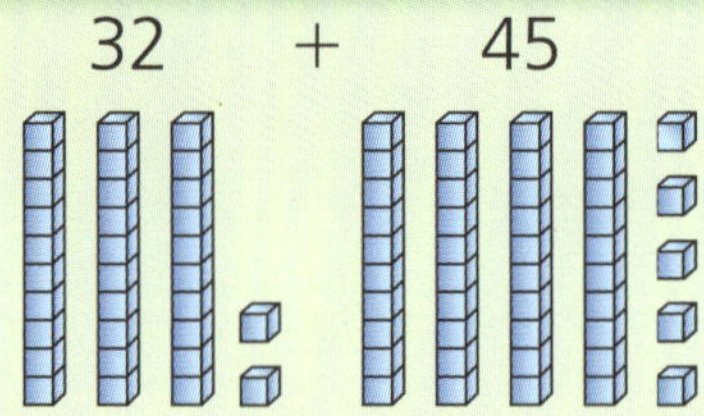

1 Use the split strategy to find the answers.

a 23 + 15
= (20 + 10) + (3 + 5)
= ☐

b 32 + 47
= (30 + 40) + (2 + 7)
= ☐

c 16 + 71
= (10 + 70) + (6 + 1)
= ☐

d 52 + 24
= ☐ + ☐
= ☐

e 81 + 15
= ☐ + ☐
= ☐

f 45 + 54
= ☐ + ☐
= ☐

2 Use the split strategy or place-value blocks to answer these.

a 14 + 14 = ☐
b 24 + 24 = ☐
c 32 + 17 = ☐
d 33 + 21 = ☐
e 18 + 51 = ☐
f 35 + 40 = ☐
g 52 + 26 = ☐
h 35 + 11 = ☐
i 22 + 66 = ☐

3 Try to do these in your head.

a 32 + 64 = ☐
b 33 + 44 = ☐
c 70 + 27 = ☐
d 52 + 30 = ☐
e 17 + 22 = ☐
f 31 + 17 = ☐
g 25 + 71 = ☐
h 24 + 81 = ☐
i 43 + 51 = ☐
j 66 + 43 = ☐
k 82 + 26 = ☐
l 35 + 24 = ☐

ICT

Use a calculator to check your work.

 • *AUSTRALIAN SIGNPOST MATHS 2* • ISBN 978 1 4886 2181 9

26B Split Strategy (Subtraction)

86 − 35

subtract tens   subtract ones

= (80 − 30) + (6 − 5)

= 51

86 − 35 = 51

1 Use the split strategy to find the answers.

a 96 − 23
= (90 − 20) + (6 − 3)
= ____

b 54 − 24
= (50 − 20) + (4 − 4)
= ____

c 56 − 35
= (50 − 30) + (6 − 5)
= ____

d 67 − 35
= ____ + ____
= ____

e 77 − 62
= ____ + ____
= ____

f 89 − 58
= ____ + ____
= ____

g 75 − 45
= ____ + ____
= ____

h 87 − 81
= ____ + ____
= ____

i 68 − 8
= ____ + ____
= ____

2 Use the split strategy or place-value blocks to answer these.

a 58 − 16 = ____
b 34 − 21 = ____
c 79 − 23 = ____
d 67 − 25 = ____
e 93 − 21 = ____
f 75 − 64 = ____

3 Try to do these in your head.

a 46 − 16 = ____
b 36 − 32 = ____
c 73 − 21 = ____
d 66 − 52 = ____
e 97 − 26 = ____
f 84 − 34 = ____

ICT: Use a calculator to check your work.

26C Seasons

Summer	Autumn	Winter	Spring
December January February	March April May	June July August	September October November

1 a The season after winter is ______.

b The season before autumn is ______.

c The first month of spring is ______.

d The first month of summer is ______.

2 Would you use minutes, hours, days or months to measure:

a the time to boil an egg? ______

b the time to build a house? ______

c the time to sleep at night? ______

In Australia, we use the same four seasons as the northern hemisphere.

Some Indigenous people in Arnhem Land, Northern Territory, use six major seasons: **Dhuludur**, **Barramirri**, **Mayaltha**, **Midawarr**, **Dharratharramirri** and **Rarrandharr**.

Use the internet to investigate these six seasons.

26D Months and Seasons

There are four seasons in one year.

1 Write the months in their correct order.

March	July	November	January
August	June	December	February
May	September	April	October

1		2	
3		4	
5		6	
7		8	
9		10	
11		12	

2 Colour your answers from Question 1 so that:

a the summer months are yellow.

b the autumn months are brown.

c the winter months are blue.

d the spring months are green.

Shopping

1 Trace or draw the coins you could use to buy each object.

a

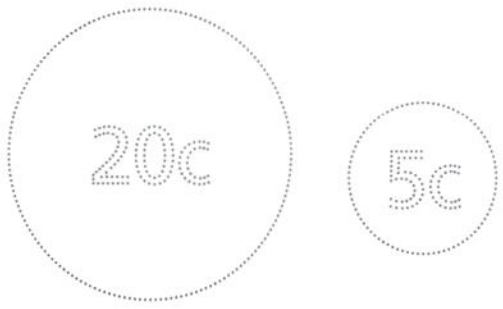

25c

b

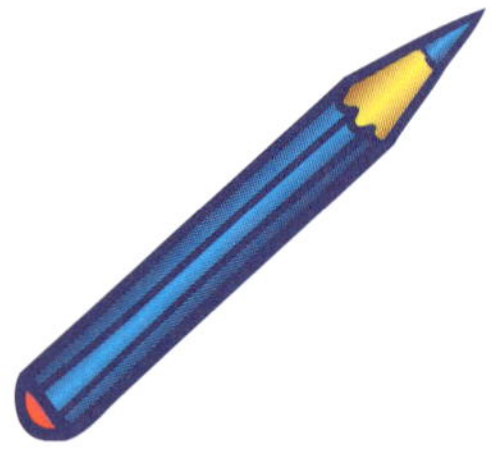

30c

c

55c

d

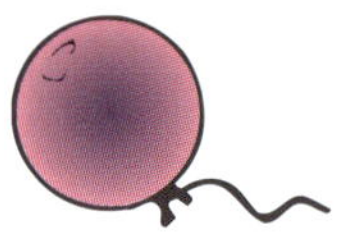

40c

e

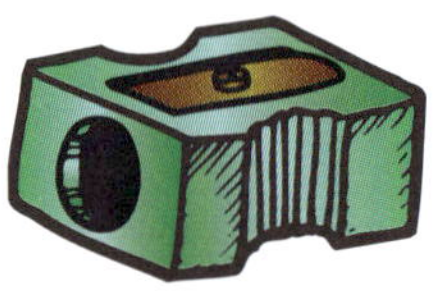

45c

f

70c

g

60c

h

65c

 • *AUSTRALIAN SIGNPOST MATHS 2* • ISBN 978 1 4886 2181 9

27B Chance

1 Is it **certain**, **likely**, **unlikely** or **impossible** that a red ball would be chosen?

2 Draw 10 apples. Colour them red or green so that green is more likely to be chosen if a friend chose without looking.

3 Roll a dice six times. Write each number rolled.

a Which number was rolled most?

b Which numbers were not rolled?

c Discuss the numbers that could be rolled.

- Choose a number on the dice.
- Count how many rolls it takes until your number is thrown. ____ rolls

27C 3D Objects

1 Write the name of each 3D object.

a

b
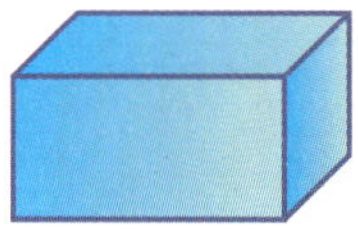

c

d
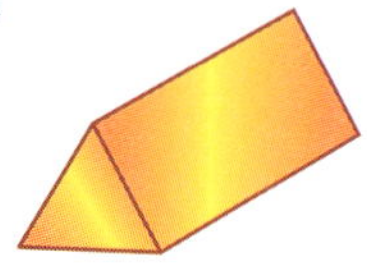

e
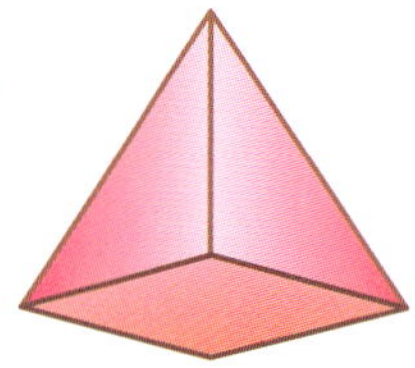

2 Which 3D objects have:

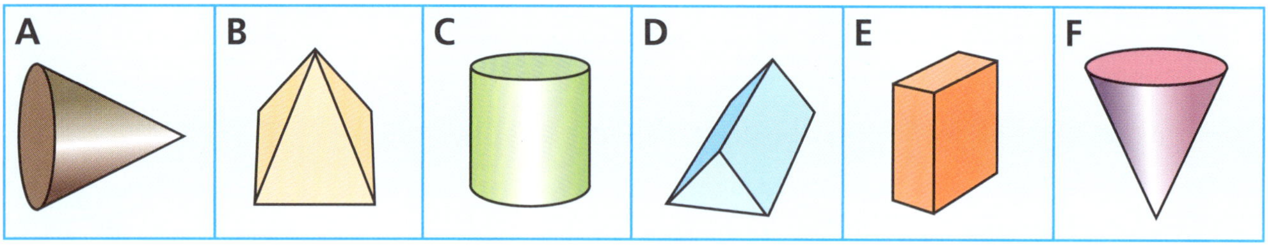

a curved surfaces?

b only flat surfaces?

c both curved and flat surfaces?

3 Which two 3D objects fit together to make these?

a
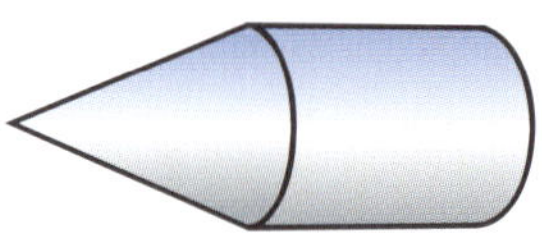

b
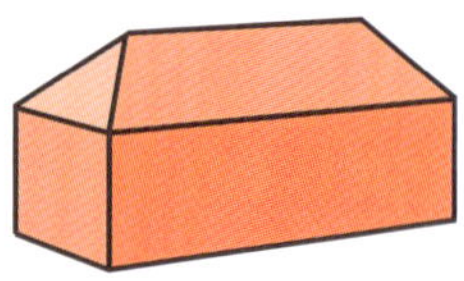

c

4 Colour cones red, cylinders blue, pyramids yellow and prisms green.

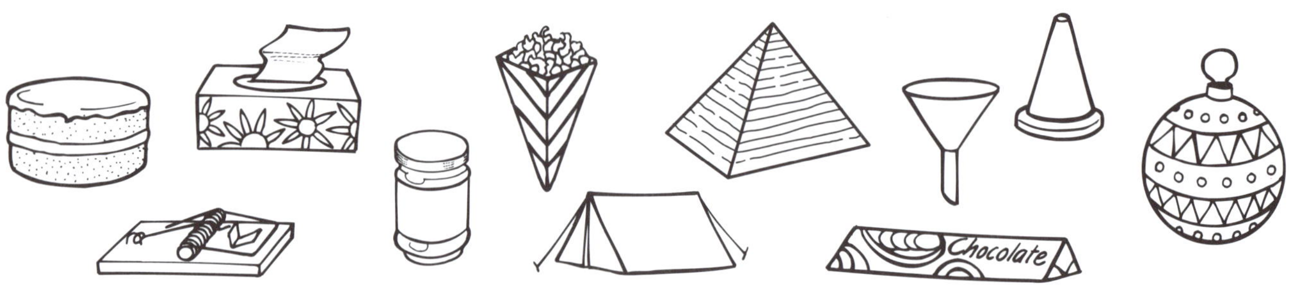

27D Drawing 2D Shapes

CONCEPT

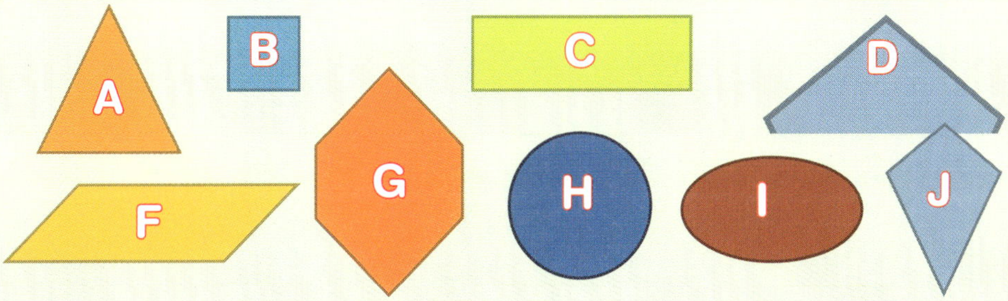

Try to name all of these shapes in order.

1 Use a ruler to draw these shapes. Use the dots to help.

a triangle **b** rectangle **c** hexagon

d rhombus **e** parallelogram **f** trapezium

ACTIVITY

Use the shapes above to make a design.

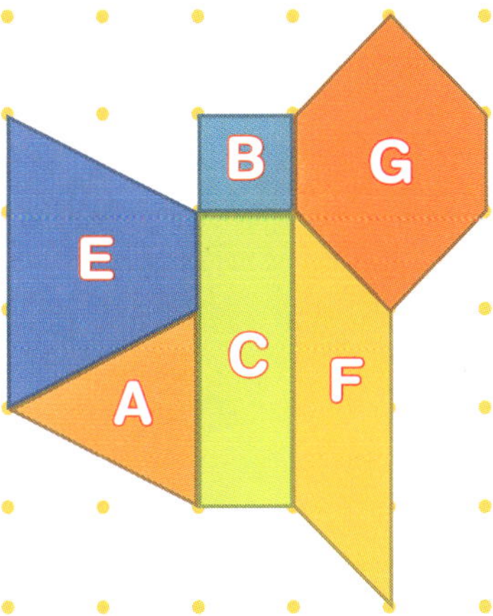

28A Division

1 Draw circles to show:

a 3 shares

How many in each share? ☐

b 6 shares

How many in each share? ☐

c 5 shares

How many in each share? ☐

d 4 shares

How many in each share? ☐

2 **a** Share 8 cakes among 4 children.

How many does each child get?

☐

b Share 12 fish between 2 tanks.

How many fish in each tank?

☐

INVESTIGATION

3 Use counters to find a fair share.

a 18 pencils shared by 6 girls = ☐ pencils

b 20 balloons shared by 5 children = ☐ balloons

c 16 children shared among 2 teams = ☐ children

 • *AUSTRALIAN SIGNPOST MATHS 2* • ISBN 978 1 4886 2181 9

28B Repeated Subtraction

We can use counters or blocks to model problems.

There are 12 apples. Put 3 in each bag.
How many bags do we need?

We can circle groups of 3.

We need **4 bags**.

1. We have 12 apples. How many people can be given:

   a 2 apples ☐ **b** 4 apples ☐ **c** 6 apples ☐

   You can use blocks or counters to do these.

2. 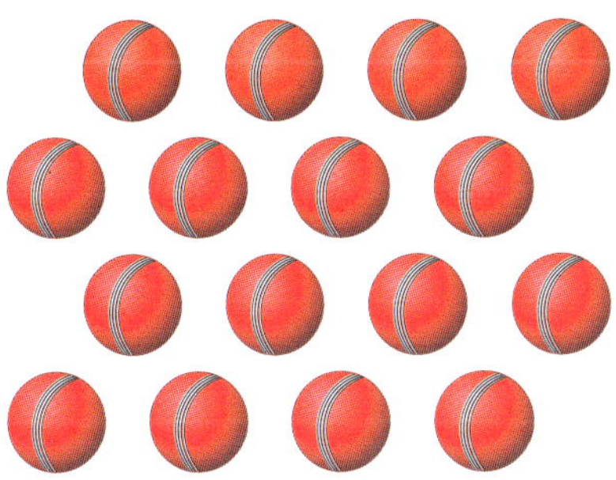

   We have 16 balls.

   How many people can be given:

   a 4 balls? ☐ **b** 8 balls? ☐

3. 

   We have 15 birds.

   How many people can be given:

   a 5 birds? ☐ **b** 3 birds? ☐

4. We have 20 lollies to share.

   How many lollies will each student get if there are:

   a 2 students? ☐ **b** 4 students? ☐

   c 5 students? ☐ **d** 10 students? ☐

 • • ISBN 978 1 4886 2181 9

28C The Calendar

Here is a calendar for October.

October						
Sun	Mon	Tues	Wed	Thurs	Fri	Sat
1	2		4	5	6	7
15	16	17	18		20	21
22	23		25	26	27	28
29		31				

Next year, October starts on a different day.

1. a Fill in the dates that are missing.
   b How many days are in October?
   c How many full weeks are there?
   d On what day does October begin?
   e On what day does October end?
   f How many Sundays are in October?
   g On what day is the 20th of October?
   h Write the long date for the last Sunday.

2. Would you use minutes, hours, days or months to measure:
   a the time it takes for a trip to the zoo?
   b the time it takes for eggs to hatch?
   c the time it takes to build a house?
   d the time it takes to cook a sausage?

28D The Calendar

CONCEPT

Thirty days has September,
April, June and November,
All the rest have thirty-one
Except February alone
Which has twenty-eight days clear,
And twenty-nine days each leap year.

1 How many days are there in these months?

a September ☐ b May ☐ c February ☐

d January ☐ e April ☐ f July ☐

March						
Sun	Mon	Tues	Wed	Thurs	Fri	Sat
		1	2	3	4	5
6	7	8	9	10	11	12
13	14	15	16	17	18	19
20	21	22	23	24	25	26
27	28	29	30	31		

2 a How many days are in March? ☐

b On what day does March begin? ☐

c On what day does March end? ☐

d How many Wednesdays are in March? ☐

e On what day is the 30th of March? ☐

f Which month follows March? ☐

g What is the date of the last Wednesday? ☐

h Circle the days of the weekends.

29A Jump Strategy (Addition)

CONCEPT

38 + 6 Start at 38 and jump to 40.

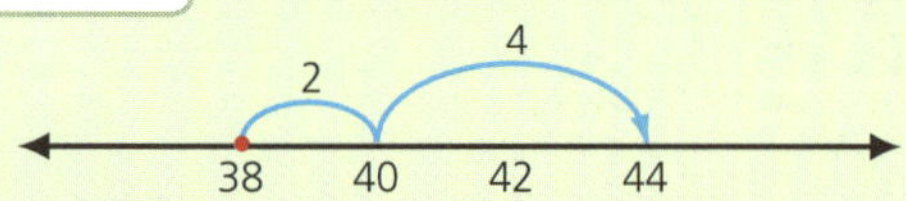

$$\begin{array}{r} 38 \\ +\ \ 6 \\ \hline 44 \end{array}$$

1 Use the number line to find the answers.

a 38 + 4 = ☐

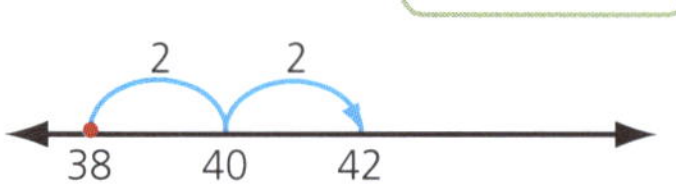

b 58 + 5 = ☐

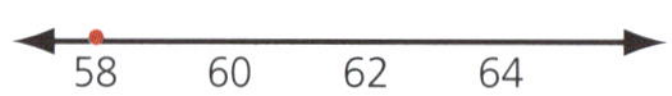

c 47 + 6 = ☐

d 79 + 3 = ☐

e 89 + 2 = ☐

f 88 + 7 = ☐

g 37 + 8 = ☐

h 57 + 9 = ☐

CONCEPT

37 + 24 Start at 37 and count on.

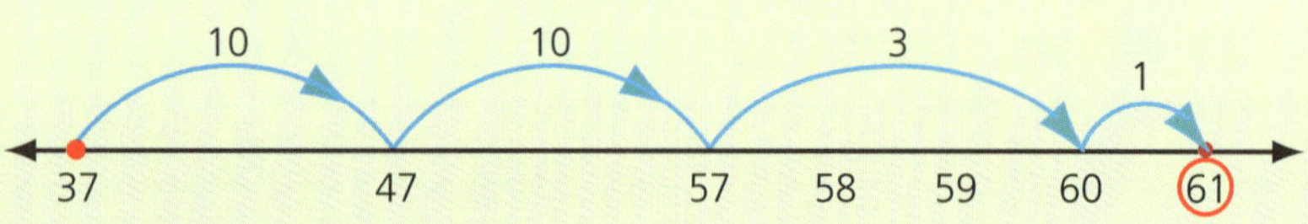

- We don't need marks on the line.
- We jump the tens, then the ones.

$$\begin{array}{r} 37 \\ +\ 24 \\ \hline 61 \end{array}$$

2 Use the jump strategy to find the answers.

a 46 + 21 = ☐

b 39 + 23 = ☐

c 58 + 14 = ☐

d 67 + 13 = ☐

29B Jump Strategy (Subtraction)

CONCEPT

63 − 24 Start at 63 and jump back.

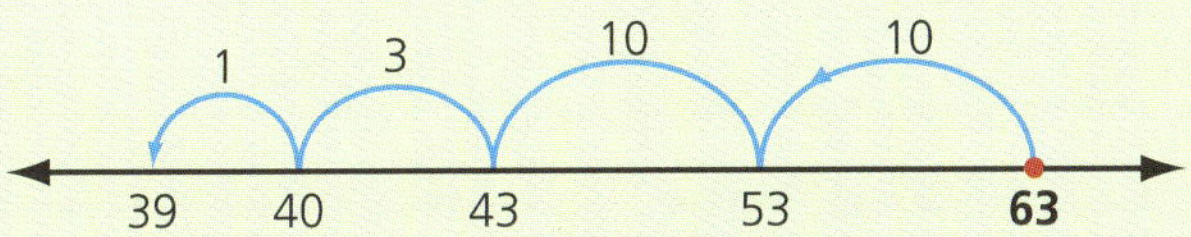

- We don't need marks on the line.
- We jump the tens, then the ones.

Subtract tens first.

63 − 24 = 39

Use the jump strategy to find the answers.

1 **a** 52 − 5 = ☐

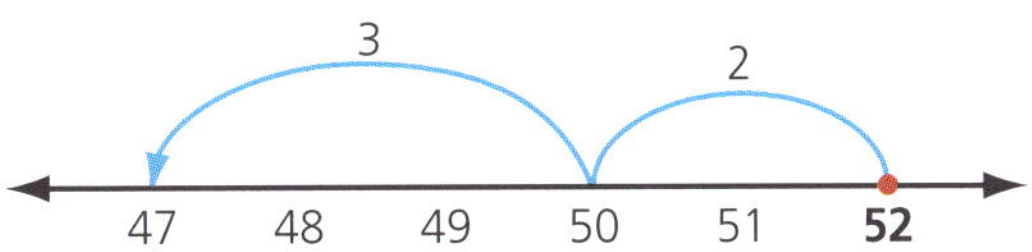

b 31 − 6 = ☐

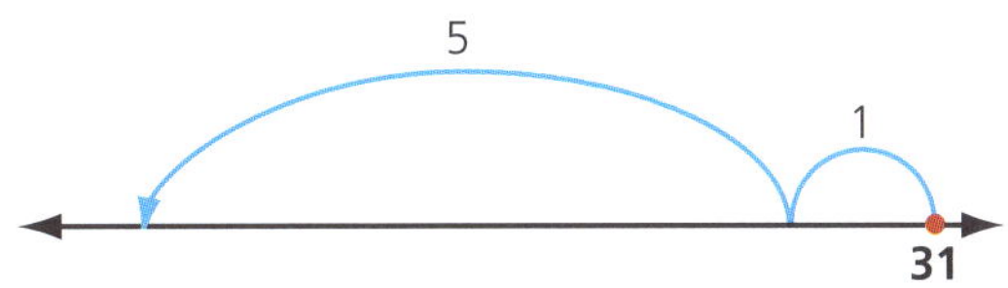

c 93 − 4 = ☐

d 71 − 3 = ☐

2 **a** 43 − 14 = ☐

b 64 − 25 = ☐

c 71 − 33 = ☐

d 82 − 17 = ☐

e 92 − 46 = ☐

f 55 − 18 = ☐

3 Can you do these in your head?

a 71 − 2 = ☐ **b** 62 − 4 = ☐ **c** 53 − 6 = ☐

d 72 − 37 = ☐ **e** 96 − 29 = ☐ **f** 64 − 46 = ☐

g 85 − 27 = ☐ **h** 31 − 17 = ☐ **i** 42 − 23 = ☐

29C Dice, Cards and Spinners

1 List all possible outcomes when a dice is rolled.

2 List all 13 types of cards that could be chosen from a pack of playing cards.

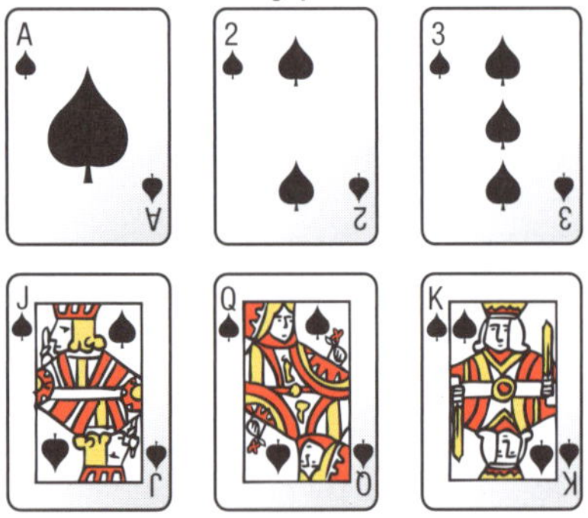

Jack Queen King

3 **a** List all possible outcomes for this spinner.

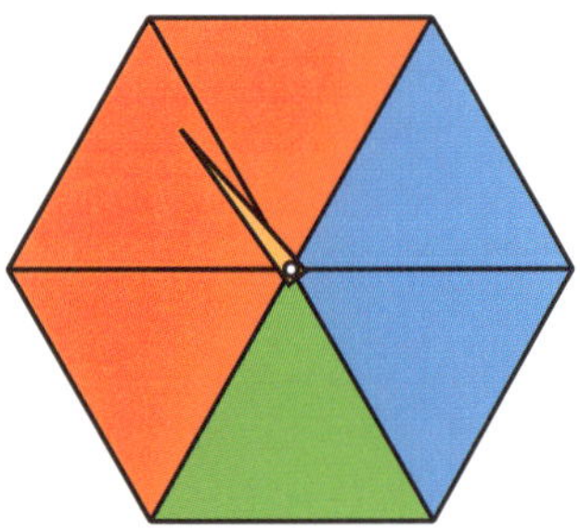

b Which colour is most likely to be spun?

c Which colour is least likely to be spun?

4 Tom chose a card from a pack of playing cards.
Is he more likely to choose:

a a picture card or a number card?

b a number from 2 to 4, or 5 to 10?

- Toss a coin until you have thrown three heads. Have a friend count the number of tosses needed. ☐ tosses
- Guess how many times it will take to toss three heads next time. ☐
  Repeat the experiment. It took ☐ tosses.

29D Following Instructions

FUN SPOT

1. Little Red Riding Hood followed these instructions. Colour her path red.

Instructions

Move
2 left
3 up
1 left
1 down
3 left
3 up
6 right
2 up
4 left

2. The man moved 4 right, then 2 down, then 4 left. Draw a tree in that square.
3. Give directions for the man to move to the wolf. Colour his path blue.
4. Give directions for the wolf to move to the bird. Colour the path green.

Make up other trips using the grid.

30A The Division Sign

CONCEPT

÷ means **shared among**
means **in groups of**
means **divided by**

$15 \div 3 = $ 5

1 Use these pictures to find **one share** if:

a 16 are shared by 4

$16 \div 4 = $ ☐

Each has ☐.

b 15 are shared by 5

$15 \div 5 = $ ☐

Each has ☐.

c 18 are shared by 9

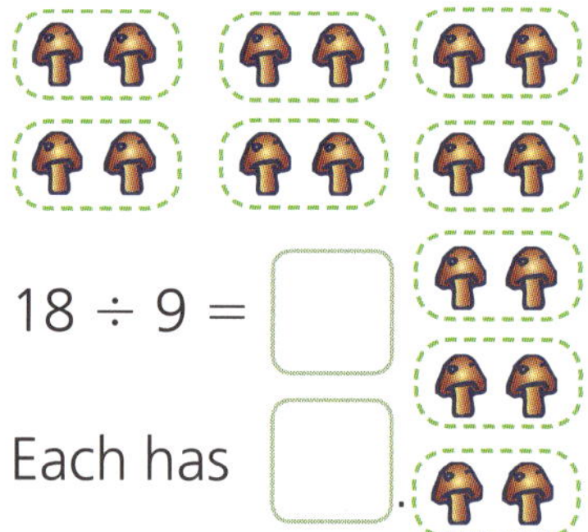

$18 \div 9 = $ ☐

Each has ☐.

d 12 are shared by 3

$12 \div 3 = $ ☐

Each has ☐.

e 20 are shared by 4

$20 \div 4 = $ ☐

Each has ☐.

f 12 are shared by 4

$12 \div 4 = $ ☐

Each has ☐.

2 Use these pictures to find out how many:

a groups of 2

$8 \div 2 = $ ☐

Each has ☐.

b groups of 6

$12 \div 6 = $ ☐

Each has ☐.

c groups of 3

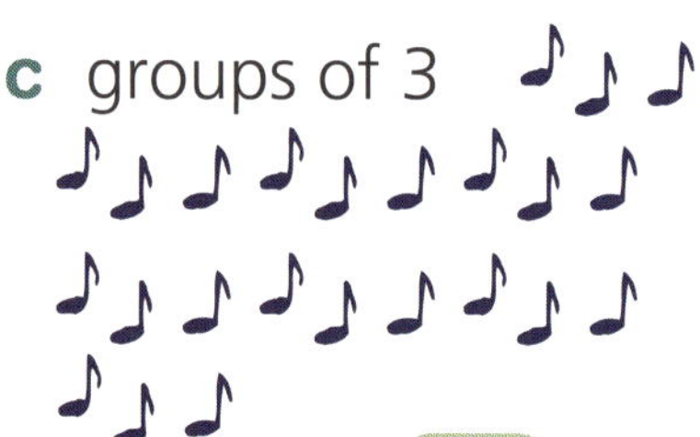

$21 \div 3 = $ ☐

Each has ☐.

d groups of 5 circles $25 \div 5 = $ ☐ There are ☐ groups.

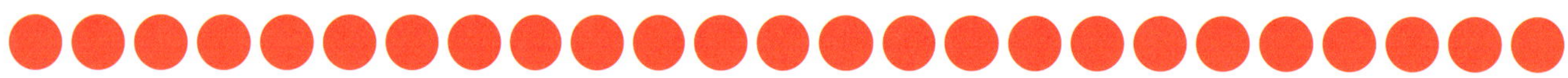

30B Division as Repeated Subtraction

CONCEPT

÷ means **shared among**
means **in groups of**
means **divided by**

How many groups of 3 are in 18?

18 ÷ 3 = 6

1

Do these by subtracting equal groups.

a How many groups of 2 in 24?
24 ÷ 2 = ☐

b How many groups of 4 in 24?
24 ÷ 4 = ☐

c How many groups of 8 in 24?
24 ÷ 8 = ☐

d How many groups of 3 in 24?
24 ÷ 3 = ☐

e How many groups of 6 in 24?
24 ÷ 6 = ☐

f How many groups of 12 in 24?
24 ÷ 12 = ☐

2

groups of 7 = ☐ 28 ÷ 7 = ☐

There are ☐ groups of 7 stars.

3

18 

a How many groups of 3 in 18? ☐ 18 ÷ 3 = ☐

b How many groups of 6 in 18? ☐ 18 ÷ 6 = ☐

c How many groups of 2 in 18? ☐ 18 ÷ 2 = ☐

Estimate before you calculate.

  ISBN 978 1 4886 2181 9

30C Comparing Areas

1 Compare the areas of these objects. List them from smallest to largest.

a	 book	 plastic lid	 envelope	**smallest** **largest**
b	 desk	 newspaper	 art paper	**smallest** **largest**

2 Compare these objects. Circle the object with the smaller area.

a clock and book
b pencil case and shoe
c towel and cupboard door
d newspaper and window
e art paper and window
f towel and table
g matchbox and book
h sheet of cardboard and window

3 Find objects in the classroom to complete the table.
Draw the object.

	Smaller Area	About the Same Area	Larger Area
newspaper 			
lid 			
book 			

30D Gathering Data

1 Rachel recorded the girls (**G**) and the boys (**B**) arriving at her party. This is the list she made.

G	G	G	B	B	B	B	G	B	G	G	G	B	B	B
G	B	B	B	G	G	B	G	B	G	B	B	G	G	B

a How many boys? ☐

b How many girls? ☐

Boys	Girls

c Make a graph to show how many boys and girls came to the party by colouring one picture for each person.

Friends at the Party

2 **a** Write true (**T**) or false (**F**) next to each sentence, then make the graph by colouring one box for each question.

$1 + 1 = 8$	F	$7 + 1 = 8$		$2 + 2 = 5$		$3 + 3 = 6$	
$1 + 5 = 6$		$2 + 5 = 9$		$2 + 3 = 5$		$5 + 1 = 6$	
$2 + 1 = 3$		$4 + 4 = 2$		$8 + 2 = 9$		$7 + 5 = 8$	
$4 + 5 = 6$		$8 + 0 = 8$		$0 + 9 = 9$		$2 + 5 = 7$	

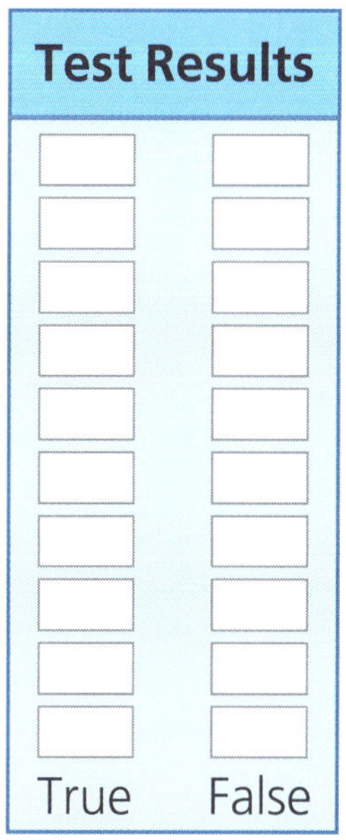

b How many were true? ☐

c How many were false? ☐

d How many questions were there altogether? ☐

 • *AUSTRALIAN SIGNPOST MATHS 2* • ISBN 978 1 4886 2181 9

31A Multiplication

CONCEPT

5 rows of 6 = 30

6 rows of 5 = 30

so

$5 \times 6 = 6 \times 5$

1

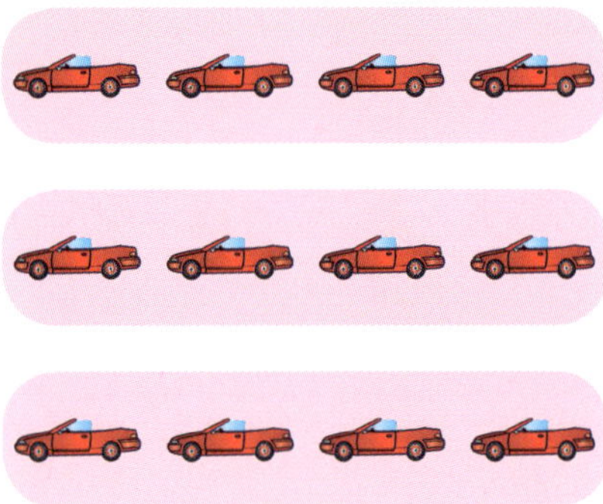

How many are in 3 rows of 4? ☐

How many are in 4 rows of 3? ☐

True or false? $3 \times 4 = 4 \times 3$ ☐

2

How many are in 3 rows of 5? ☐

How many are in 5 rows of 3? ☐

True or false? $3 \times 5 = 5 \times 3$ ☐

3

How many are in 4 rows of 10? ☐

How many are in 10 rows of 4? ☐

True or false? $4 \times 10 = 10 \times 4$ ☐

4 True or false?

$2 \times 10 = 10 \times 2$ ☐ $3 \times 6 = 6 \times 3$ ☐ $8 \times 4 = 4 \times 8$ ☐

31B Division as Repeated Subtraction

CONCEPT

We need to find out how many groups of 3 nails in 27.

$27 \div 3 = 9$

Put a dot after each group of 3. Count the groups of 3.

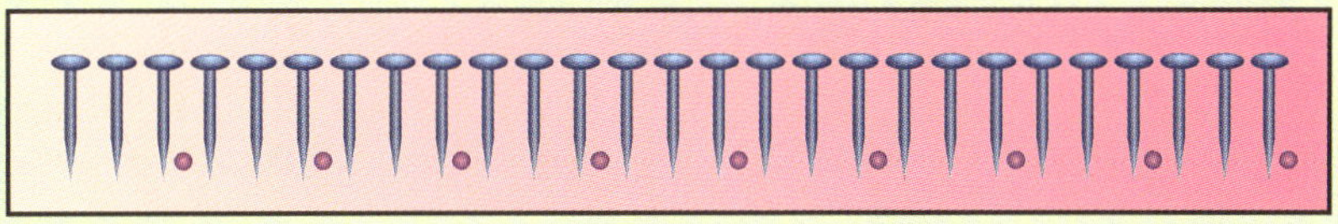

Do these by subtracting equal groups.

1

a How many groups of 2 kites are in 10? ☐ $10 \div 2 =$ ☐

b How many groups of 5 kites are in 10? ☐ $10 \div 5 =$ ☐

2

a How many groups of 6 cars are in 12? ☐ $12 \div 6 =$ ☐

b How many groups of 4 cars are in 12? ☐ $12 \div 4 =$ ☐

3

a How many groups of 2 balls are in 14? ☐ $14 \div 2 =$ ☐

b How many groups of 7 cars are in 14? ☐ $14 \div 7 =$ ☐

INVESTIGATION

4 Start with 36 counters. Use repeated subtraction for these.

a $36 \div 4 =$ ☐ **b** $36 \div 9 =$ ☐ **c** $36 \div 12 =$ ☐

d $36 \div 3 =$ ☐ **e** $36 \div 6 =$ ☐ **f** $36 \div 1 =$ ☐

g $36 \div 2 =$ ☐ **h** $36 \div 18 =$ ☐ **i** $36 \div 36 =$ ☐

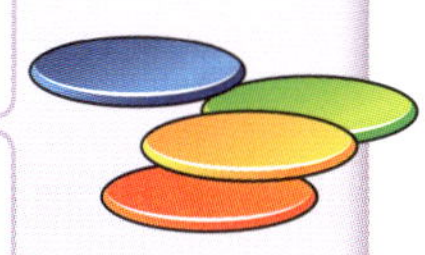

 • *AUSTRALIAN SIGNPOST MATHS 2* • ISBN 978 1 4886 2181 9

31C Turning a Shape

CONCEPT

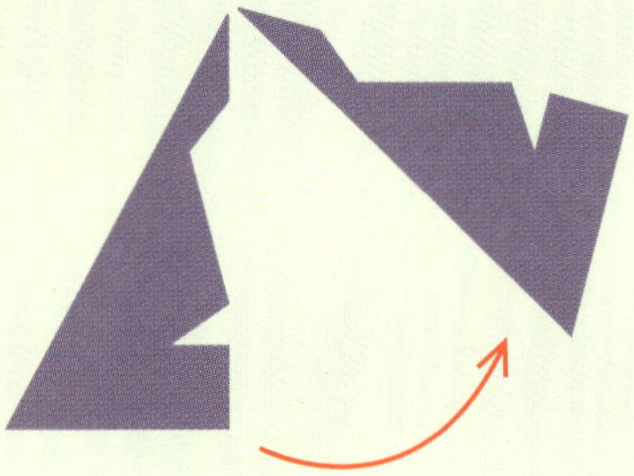

Turn the shape.

A

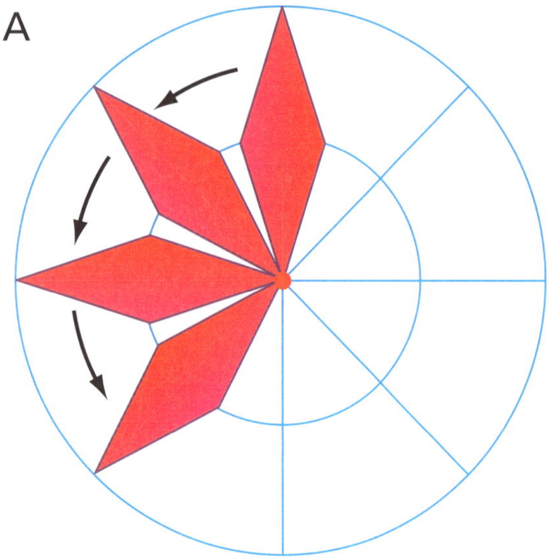

B

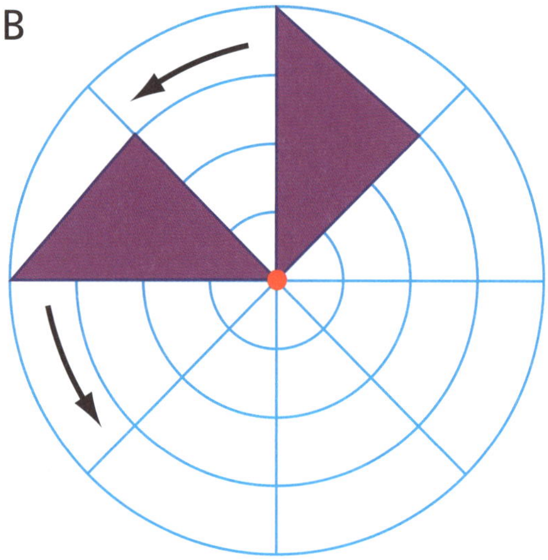

1 In picture **A** above, a pattern block has been traced, then turned, and then traced and turned again. Complete the pattern.

a Does the shape change as it turns? ☐

b Does the shape turn around a point? ☐

2 Complete pattern **B**.

ACTIVITY

- Play **Simon Says**, giving directions using the words 'slide' and 'turn'.
- 'Turn clockwise' means 'turn right'.
- 'Turn anticlockwise' means 'turn left'.

Slide 2 steps to the right. Slide 4 steps back.

Turn clockwise a half turn.

Turn a quarter turn left.

Use a computer to show flips, slides and turns.

 • *AUSTRALIAN SIGNPOST MATHS 2* • ISBN 978 1 4886 2181 9

31D Turning Shapes

1

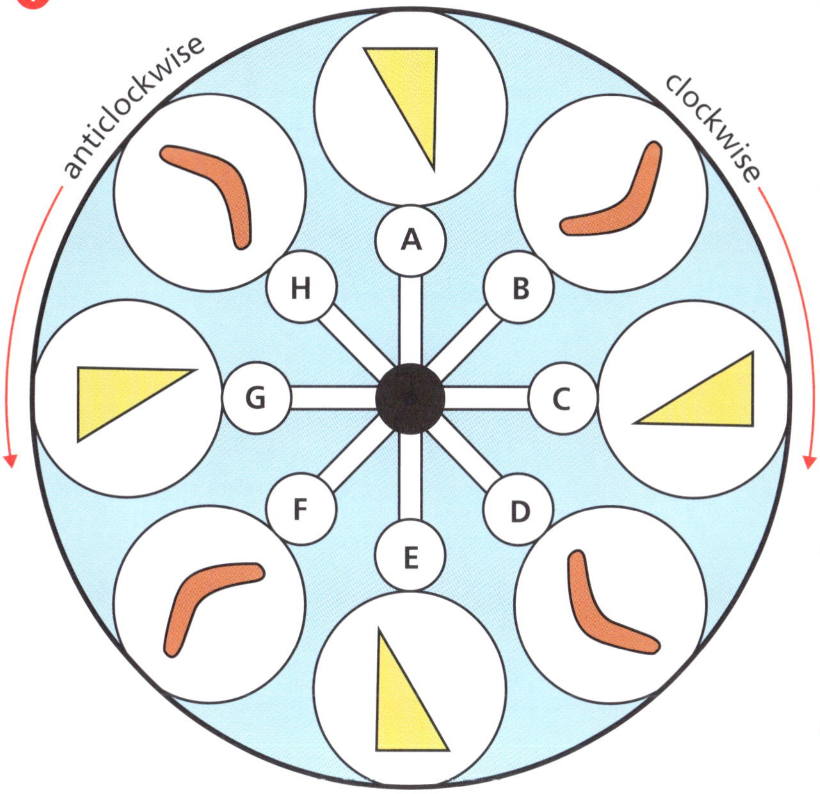

a When **A** turns a quarter turn clockwise, it becomes shape ☐.

b When **A** turns a quarter turn anticlockwise, it becomes shape ☐.

c When **A** turns a half turn clockwise, it becomes shape ☐.

d When **B** turns a half turn clockwise, it becomes shape ☐.

e When **B** turns a quarter turn clockwise, it becomes shape ☐.

Using Pattern Blocks

ACTIVITY

2 a Use a pattern block to finish this pattern. Turn it around the centre and trace the block.

b Make other turning patterns using the pattern blocks.

3 Use pattern blocks to make patterns by flipping and sliding.

32A Subtraction Using Blocks

CONCEPT

56 − 23 = 33

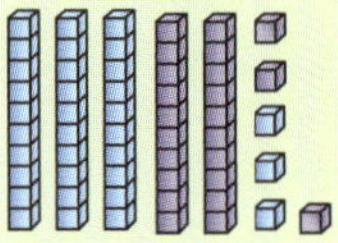

31 − 19 = 12

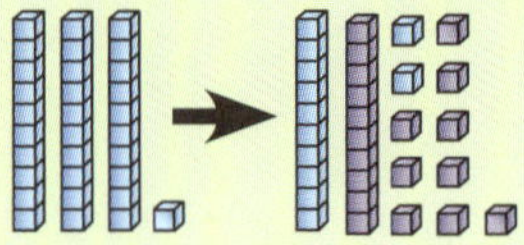

1 Write the number modelled.

a 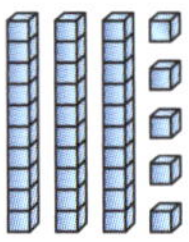 ☐

b 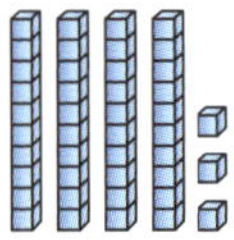 ☐

c 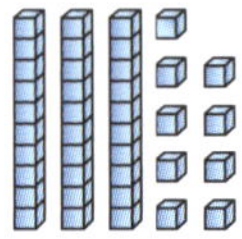 ☐

d 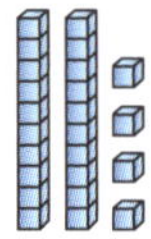 ☐

ICT

2 Complete each of these. Check your answers using a calculator.

a 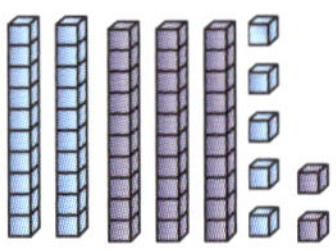

57 − 32 = ☐

b 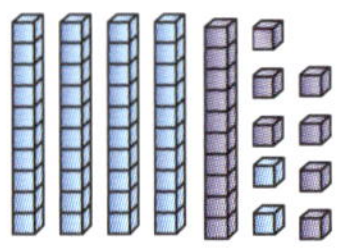

59 − 17 = ☐

c 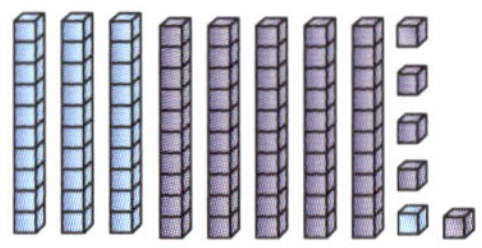

86 − 55 = ☐

d 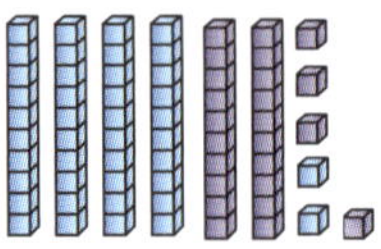

66 − 24 = ☐

e 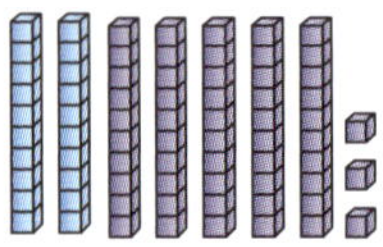

73 − 53 = ☐

f 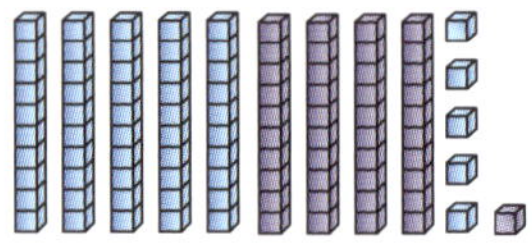

96 − 41 = ☐

g 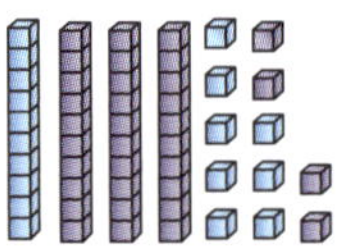

52 − 34 = ☐

h 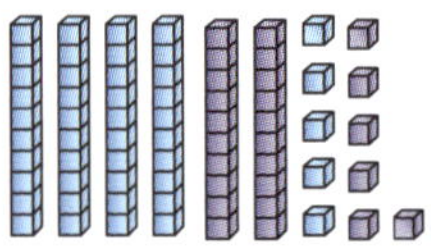

71 − 26 = ☐

i 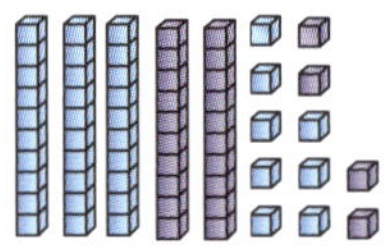

62 − 24 = ☐

j 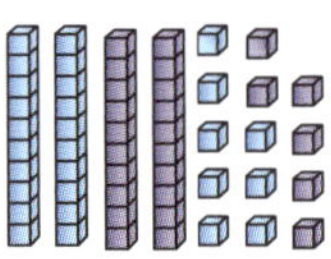

54 − 26 = ☐

k 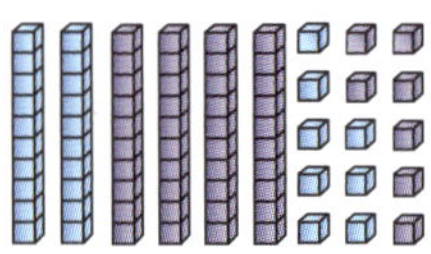

75 − 47 = ☐

l 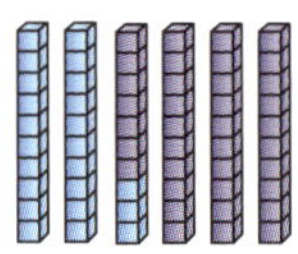

60 − 37 = ☐

32B Addition Using Blocks

CONCEPT

33 + 26 = 59

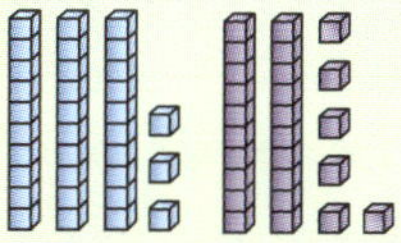

Here I have 5 tens and 9 ones.

Here I have 3 tens and 12 ones. That's the same as 4 tens and 2 ones.

Building to ten.

13 + 29 = 42

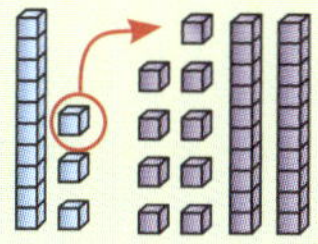

1 Write the number modelled.

a 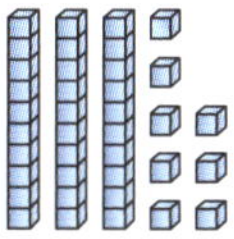 ☐

b 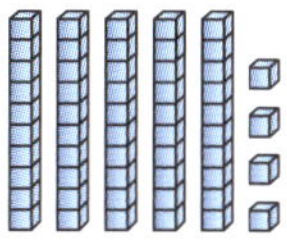 ☐

c 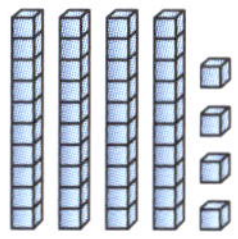 ☐

d 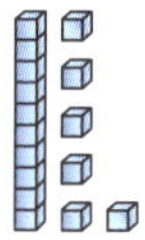 ☐

ICT

2 Complete each of these. Check your answers using a calculator.

a 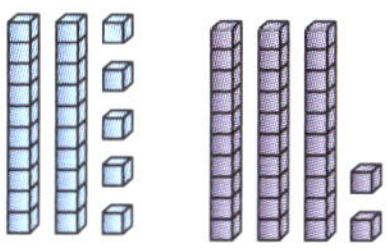

25 + 32 = ☐

b 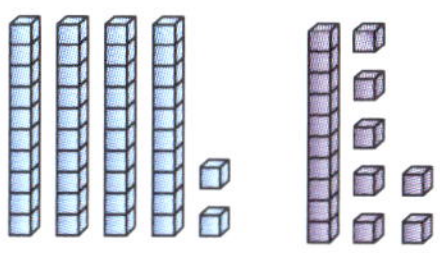

42 + 17 = ☐

c 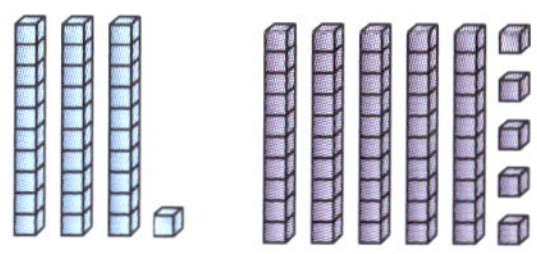

31 + 55 = ☐

d 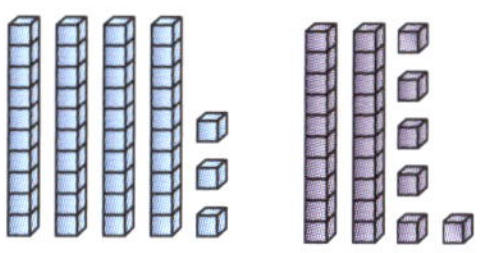

43 + 26 = ☐

e 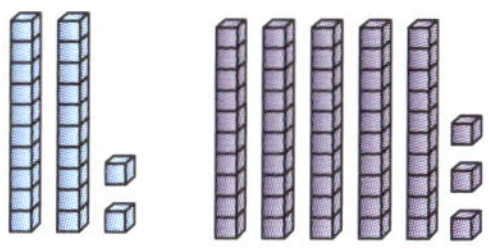

22 + 53 = ☐

f 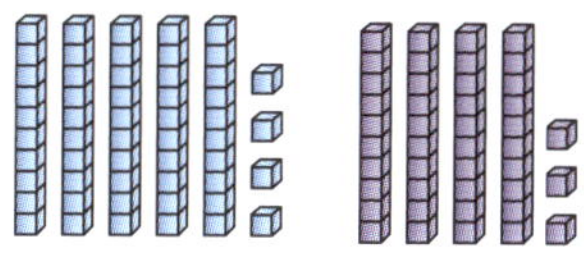

54 + 43 = ☐

g 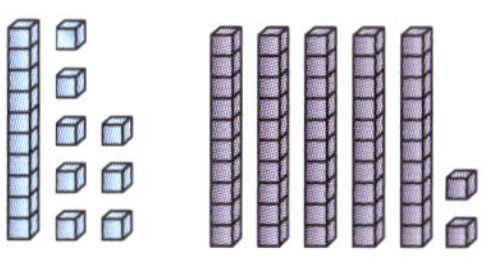

18 + 52 = ☐

h 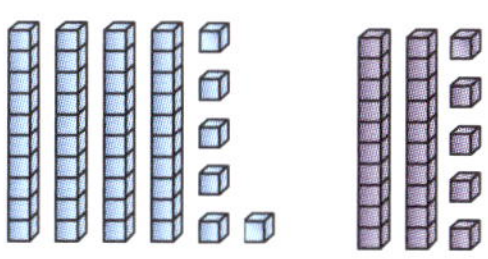

46 + 25 = ☐

i 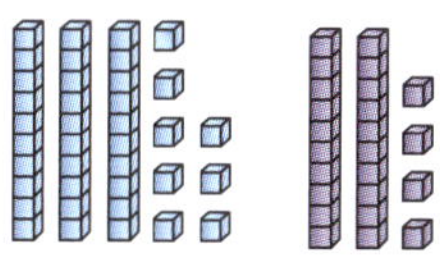

38 + 24 = ☐

j 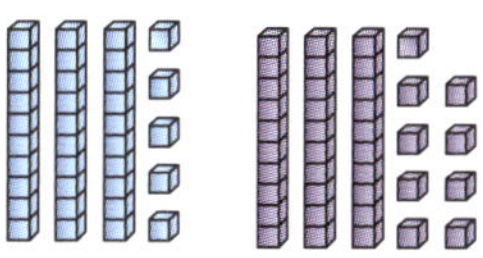

35 + 39 = ☐

k 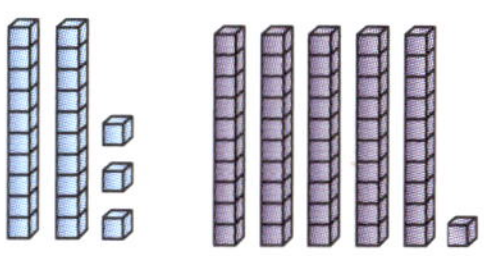

23 + 51 = ☐

l 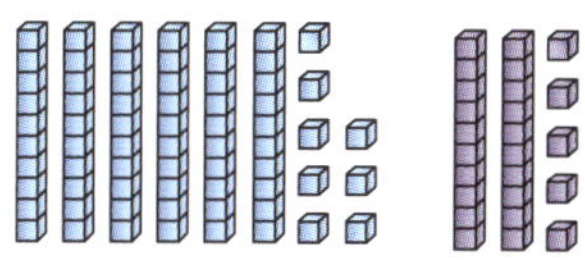

68 + 25 = ☐

 • *AUSTRALIAN SIGNPOST MATHS 2* • ISBN 978 1 4886 2181 9

32C Quarters and Eighths

CONCEPT

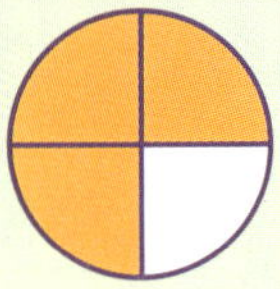

$\frac{3}{4}$ is coloured.

3 of 4 equal parts

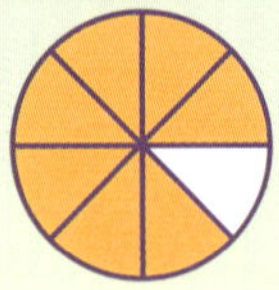

$\frac{7}{8}$ is coloured.

7 of 8 equal parts

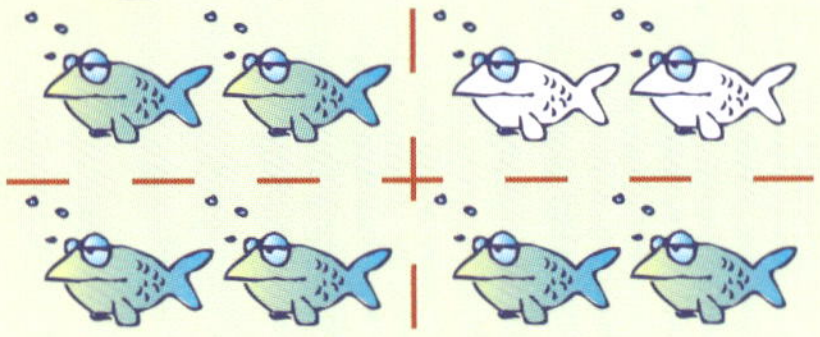

quarters

eighths

1 What part of each group is coloured?

a 

b 

c 

d 

e 

f 

g 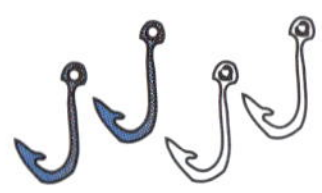

h 

2 Colour part of each group to match the fraction.

a 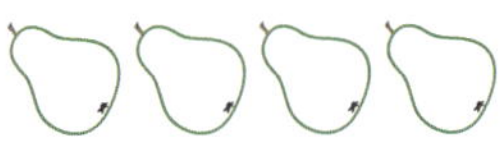 $\frac{2}{4}$

b 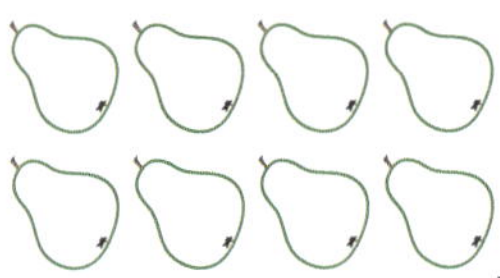 $\frac{5}{8}$

c  $\frac{3}{8}$

d  $\frac{3}{4}$

e 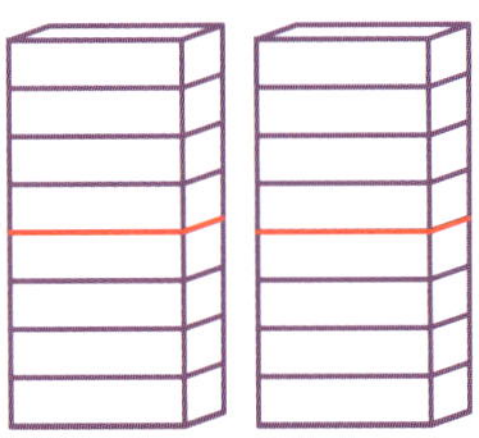 $\frac{3}{4}$

f 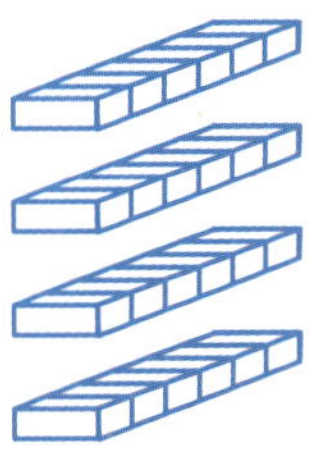 $\frac{1}{4}$

g  $\frac{3}{4}$

h 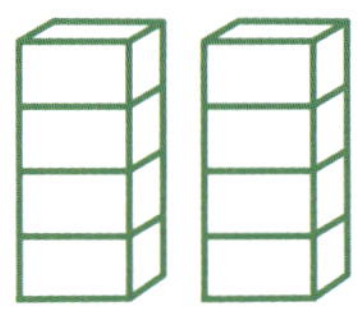 $\frac{5}{8}$

ACTIVITY

The fraction coloured is ____________.

Use counters to model groups split into quarters and eighths.

Use them to show $\frac{3}{4}$ and $\frac{3}{8}$.

32D Naming Fractions

CONCEPT

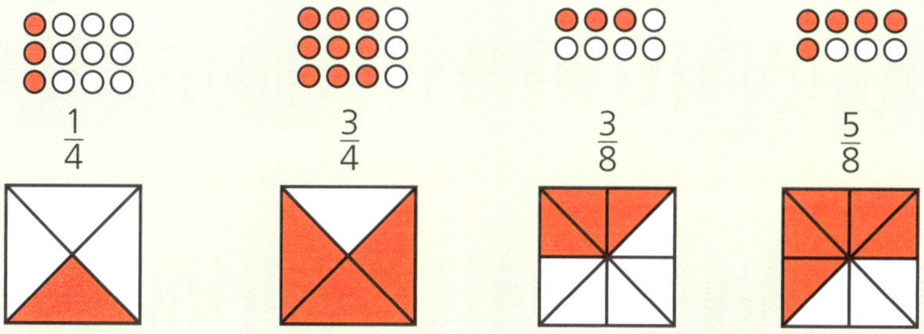

$\frac{1}{4}$ one quarter

$\frac{3}{4}$ three quarters

$\frac{3}{8}$ three eighths

$\frac{5}{8}$ five eighths

1 Colour part of each shape to match the fraction.

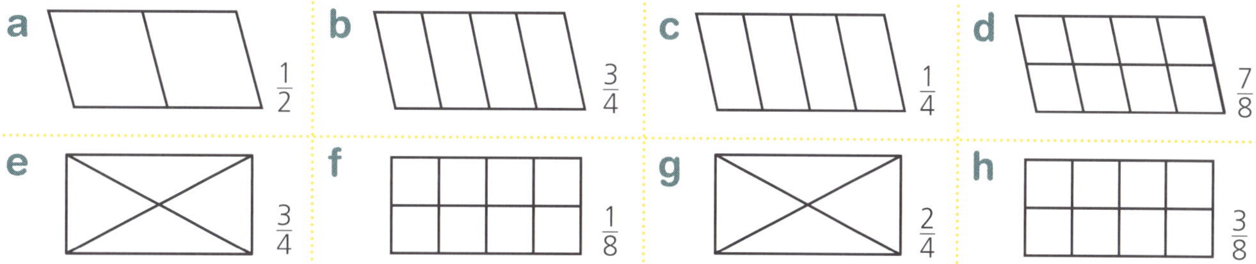

a $\frac{1}{2}$ b $\frac{3}{4}$ c $\frac{1}{4}$ d $\frac{7}{8}$

e $\frac{3}{4}$ f $\frac{1}{8}$ g $\frac{2}{4}$ h $\frac{3}{8}$

2 What part of each shape has been coloured: $\frac{1}{2}$, $\frac{1}{4}$, $\frac{3}{4}$, $\frac{1}{8}$, $\frac{3}{8}$ or $\frac{5}{8}$?

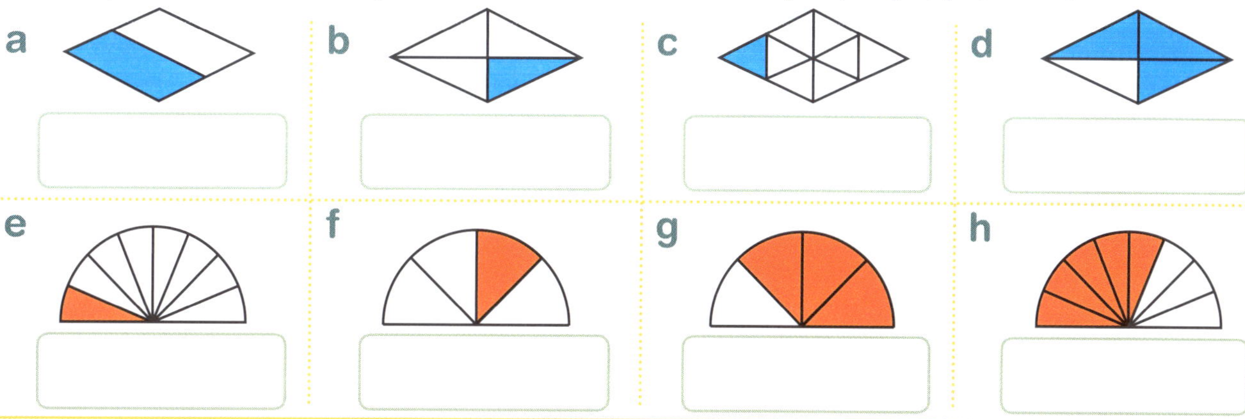

a b c d

e f g h

3 What part of each group has been coloured?

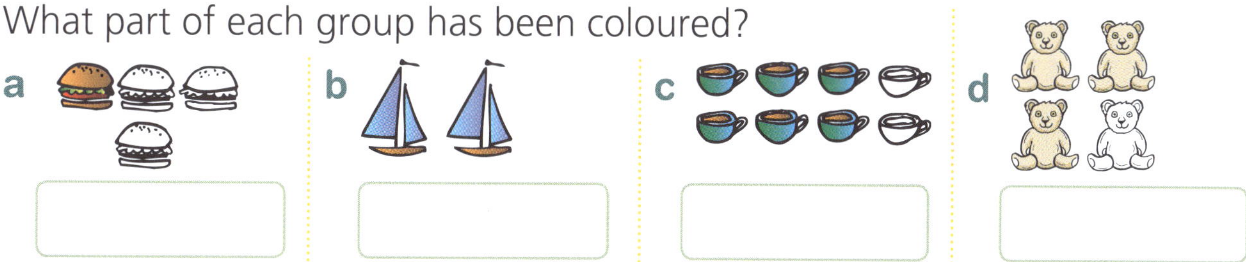

a b c d

4 Colour part of each group to match the fraction.

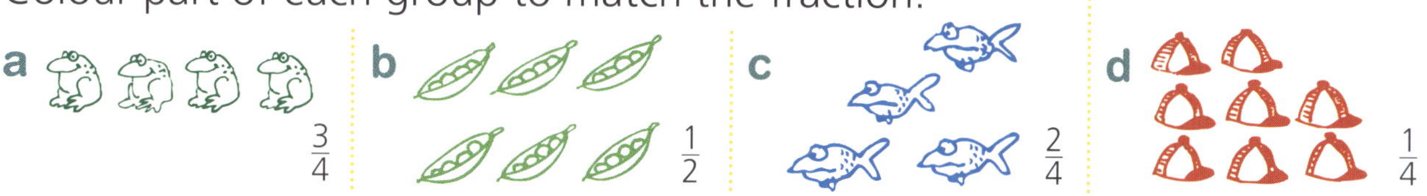

a $\frac{3}{4}$ b $\frac{1}{2}$ c $\frac{2}{4}$ d $\frac{1}{4}$

5 Is one half the same as two quarters? Discuss why or why not.

33A Choosing a Strategy

1. Match each question with the most appropriate strategy. Explain why you chose each strategy.

Strategy	Question	Strategy
doubles	36 + 10	near doubles
	57 − 12	
split strategy	22 + 3	jump strategy
	6 + 7	
adding 10	19 − 3	counting on
	42 − 16	
10 facts	17 + 3	counting back

2. Write the best strategies to solve each problem. Discuss why you chose each strategy.

a 64 − 10 = ☐ subtracting 10 or using blocks

b 39 + 4 = ☐

c 9 + 9 = ☐

d 29 + 34 = ☐

e 7 + 8 = ☐

f 64 − 23 = ☐

g 27 + 3 = ☐

h 18 − 9 = ☐

3. Use a calculator to answer each of these.

a 14 + 69 = ☐

b 83 − 58 = ☐

c 24 + 77 = ☐

d 94 − 69 = ☐

 • *AUSTRALIAN SIGNPOST MATHS 2* • ISBN 978 1 4886 2181 9

33B Choosing a Strategy

a 39 − 12 = 27

(30 − 10) + (9 − 2)

= 20 + 7

= 27

I used the split strategy.
I subtracted the tens, then the ones.

b 49 + 14 = 63

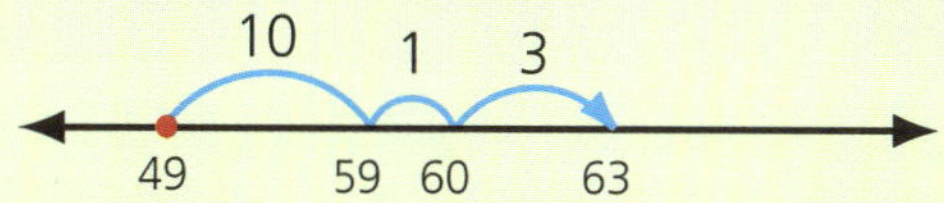

I used the jump strategy.
I added 10, then 1, then 3.

1 Use the boxes to show how you found each answer.

a 8 + 9 = ☐

b 56 − 10 = ☐

c 36 + 4 = ☐

d 41 − 4 = ☐

e 28 + 16 = ☐

f 72 + 15 = ☐

 • *AUSTRALIAN SIGNPOST MATHS 2* • ISBN 978 1 4886 2181 9

33C Faces, Edges and Corners

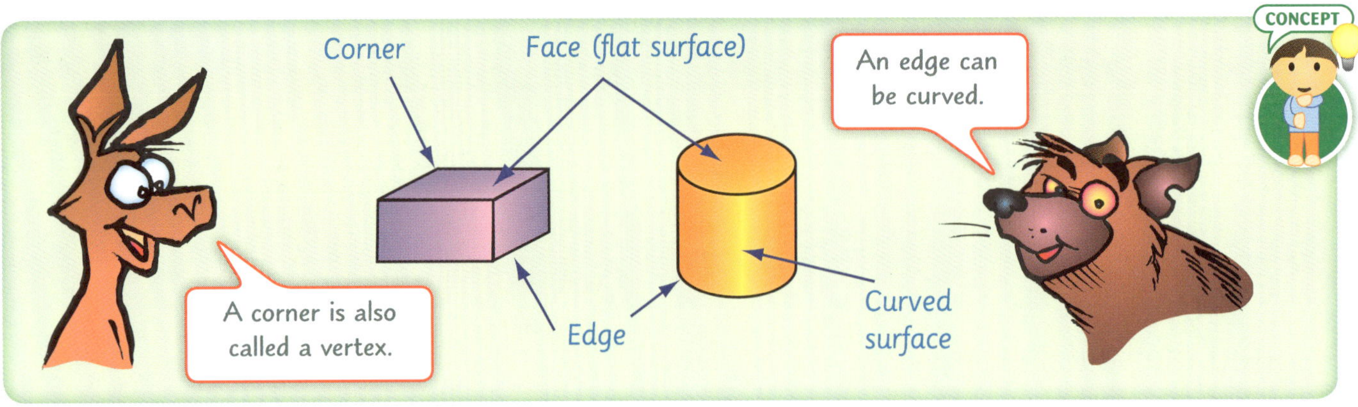

❶ Find objects like these and complete the table.

Object	Number of Faces	Number of Edges	Number of Corners	Number of Curved Surfaces
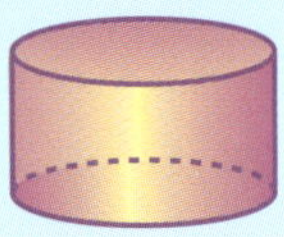				
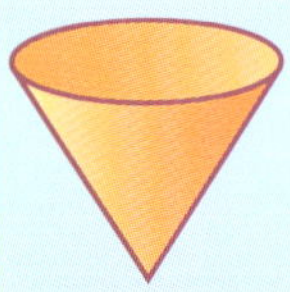				
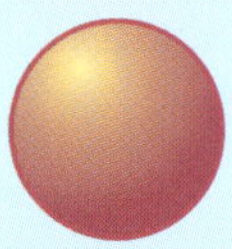				
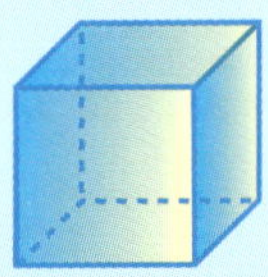				
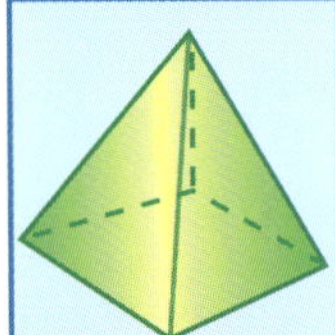				

An edge is where two surfaces meet.

 • *AUSTRALIAN SIGNPOST MATHS 2* • ISBN 978 1 4886 2181 9

33D Comparing Volume

Use blocks to build these models.

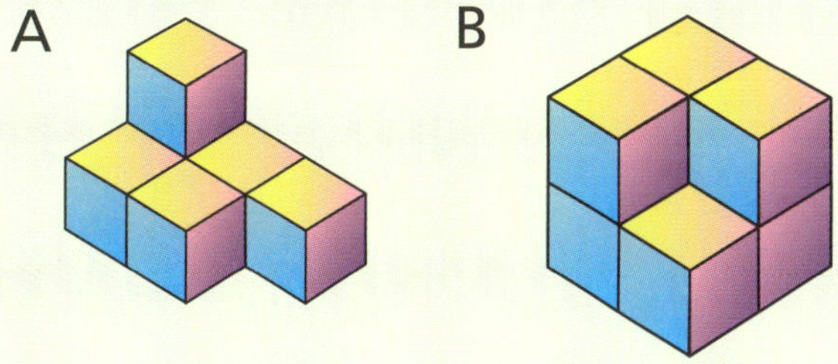

1 a How many blocks in A? ☐

b How many blocks in B? ☐

c Which model contains more blocks? ☐

d Which model has the larger volume? ☐

e Which model has the smaller volume? ☐

2 Count the number of blocks in each model.

a ☐ b ☐ c ☐ d ☐

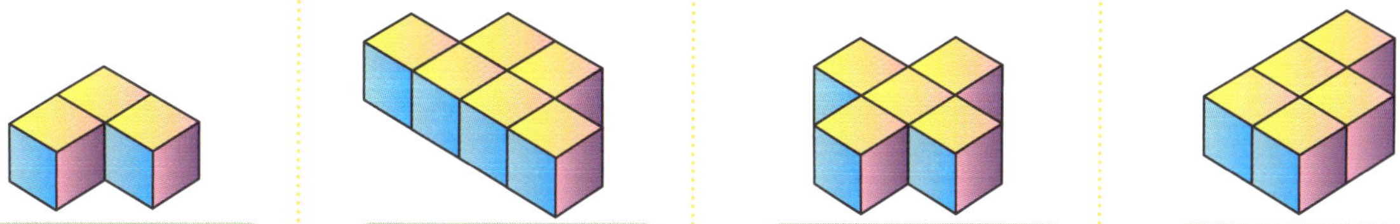

e ☐ f ☐ g ☐ h ☐

i ☐ j ☐

3 a Draw lines to join models that have the same volume.

b Circle the model with the greatest volume.

c Tick the model with the least volume.

34A Problem Solving

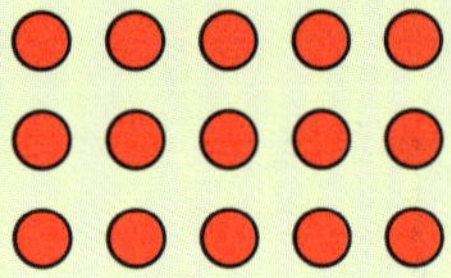

3 rows of 5 = 15
$3 \times 5 = 15$

The number of 5s in 15 = 3
$15 \div 5 = 3$

Multiplication is related to division.

3 rows make 15, so there are 5 in each row.
$15 \div 3 = 5$

1 **a** How many oranges are in three bags with four oranges in each bag?

b If there are 12 oranges and there are 4 in each bag, how many bags?

c If there are 12 oranges shared between 3 bags, how many in each bag?

2 A can of tennis balls holds 3 balls. How many balls could be placed in 5 cans?

3 How many groups of 3 could be made from a class of 30 students?

4 A class of 30 students was placed into 3 groups. How many were in each group?

Working

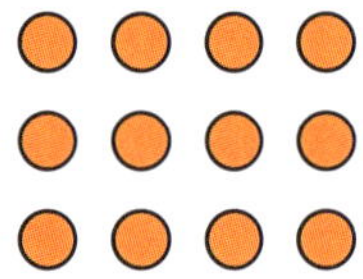

Draw as many arrays as you can that use 24 tiles.
This array is 12 squares long and 2 squares wide.

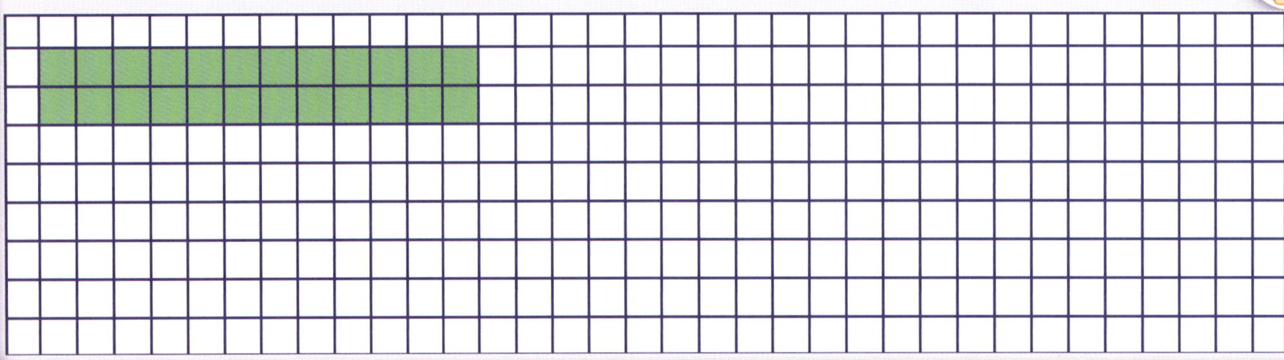

 • *AUSTRALIAN SIGNPOST MATHS 2* • ISBN 978 1 4886 2181 9

34B Counting Coins

An amount of money can often be made up in many different ways.

In what other ways could we make $1?

$1

$1

$1

A

B

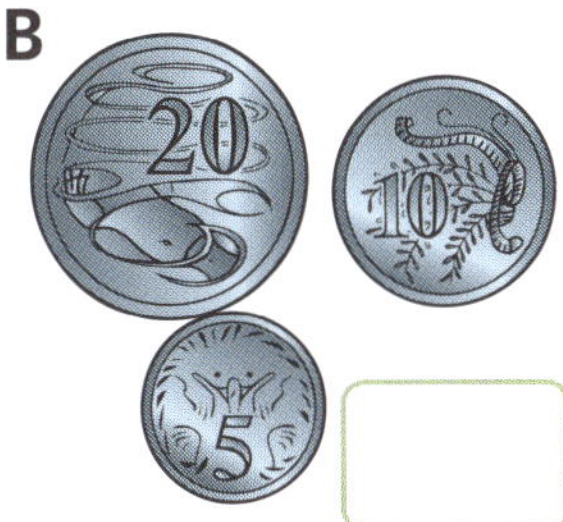

C

D

E

F

G

H

1 Which group of coins is:

a worth the same as **A**? ☐ b worth the same as **F**? ☐

c worth less than **B**? ☐ d worth more than **G**? ☐

2 What is the total value of the coins in:

a **A** and **D**? ☐ b **E** and **F**? ☐ c **C** and **G**? ☐

3 Choose any three coins. Write their total value.
For example, ($2) (20c) (10c) = $2.30

a ◯ ◯ ◯ = ☐ b ◯ ◯ ◯ = ☐

c ◯ ◯ ◯ = ☐ d ◯ ◯ ◯ = ☐

 • *AUSTRALIAN SIGNPOST MATHS 2* • ISBN 978 1 4886 2181 9

34C Informal Units of Length

How many times will the width of my finger fit along my pencil?

CONCEPT

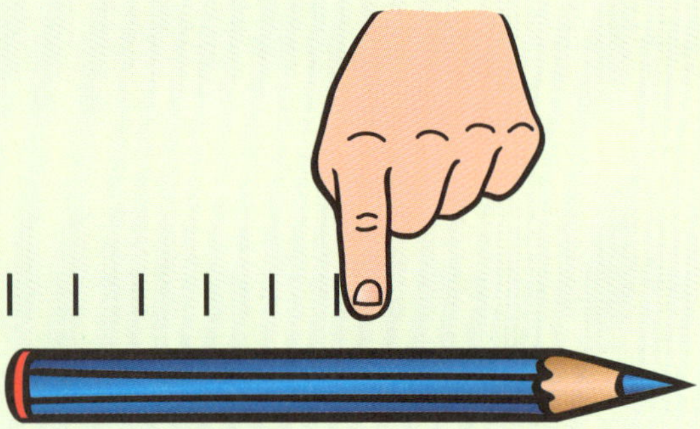

1 Use the object named as a measuring unit to find the length of your desk.

Unit Used	Length of Desk
this book	☐ books
your pen	☐ pens
finger lengths	☐ finger lengths

ACTIVITY

2 Estimate then measure how many steps to:

	Guess	Check
a the school canteen		
b the lunch seats		
c the library.		

Order these distances from shortest to longest.

3 Use hand spans to measure the length of:

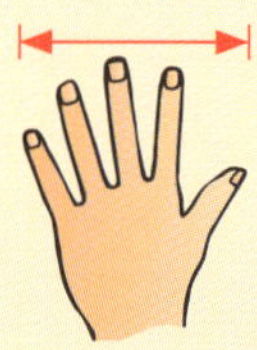

	Guess	Check
a this book		
b your desk		
c a window		
d your arm.		

Order these lengths from shortest to longest.

34D Comparing Objects

An estimate is your best guess. Estimate the answers to Questions 1 and 2.

Rocks

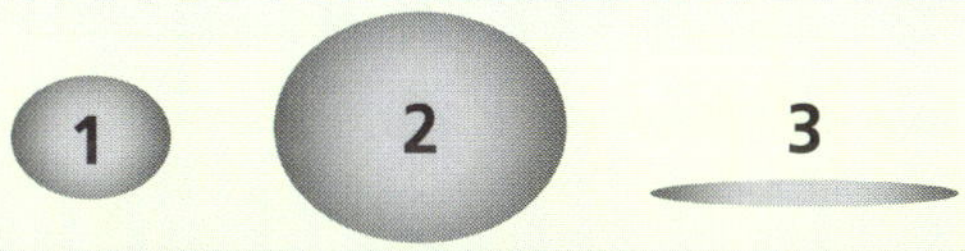

1. Which of these rocks is:
   - a the tallest? ☐
   - b the shortest? ☐
   - c the heaviest? ☐
   - d the lightest? ☐
   - e the widest? ☐
   - f the narrowest? ☐
   - g the one that takes up the most space (has the largest volume)? ☐
   - h the one that takes up the least space (has the smallest volume)? ☐

Containers

2. Which of these containers:
   - a holds the most water? ☐
   - b is the lightest? ☐
   - c is the widest? ☐
   - d holds the least water? ☐
   - e is the heaviest? ☐
   - f is the narrowest? ☐
   - g has the most outside area? ☐

3. Find three objects and draw them. Compare them. Draw lines to match each object with its properties.

Heft the objects to find which is heavier.

Tallest | Shortest | Biggest area | Smallest area | Heaviest | Lightest

35A Using a Calculator

How many balls would be in 37 cans?

37 × 3 =

There are 111 balls.

We have 123 balls in cans. How many cans?

123 ÷ 3 =

There are 41 cans.

CONCEPT

There are 3 balls in a can.

1 Write the calculator sentence for each.

a 3 balls are in each can. ☐ ☐ ☐ ☐

How many balls in 28 cans? There are ☐ balls.

b We have 108 balls in cans of 3. ☐ ☐ ☐ ☐

How many cans do we have? There are ☐ cans.

c 15 packets hold 90 ice creams. ☐ ☐ ☐ ☐

How many ice creams are in each packet? There are ☐ in each.

d 8 packets of lollies hold 128 lollies altogether. How many are in each packet? ☐ ☐ ☐ ☐

There are ☐ in each.

e How many bags of 5 apples are there if we have 125 apples? ☐ ☐ ☐ ☐

f How many apples are in each bag if 114 are shared between 19 bags? ☐ ☐ ☐ ☐

There are ☐ in each bag.

ICT

2 Make up your own calculator problems.

☐ clowns can fit into each car.

a How many clowns fit in ☐ cars? ☐

b How many clowns fit in ☐ cars? ☐

 • *AUSTRALIAN SIGNPOST MATHS 2* • ISBN 978 1 4886 2181 9

35B The Calculator

Look at your calculator to find:

- the [+] key
- the [−] key
- the [×] key
- the [=] key

1 Try to predict the answer, then use a calculator to complete these. Describe the patterns you see.

a 789 [+] 2 [=] [=] [=] [=] [=] [=]

Display shows	791					

b 456 [+] 5 [=] [=] [=] [=] [=] [=]

Display shows	461					

c 123 [+] 10 [=] [=] [=] [=] [=] [=]

Display shows	133					

d 741 [+] 5 [=] [=] [=] [=] [=] [=]

Display shows	746					

2 Press these numbers, then turn your calculator upside down. Write what you see.

a 335 ______

b 7735 ______

c 3504 ______

 • *AUSTRALIAN SIGNPOST MATHS 2* • ISBN 978 1 4886 2181 9

35C Flip, Slide and Turn

CONCEPT

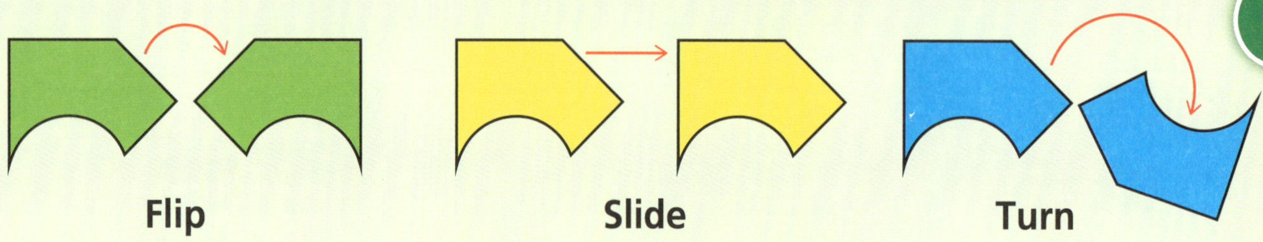

1. Continue these patterns by flipping or sliding the coloured shape.

a

b

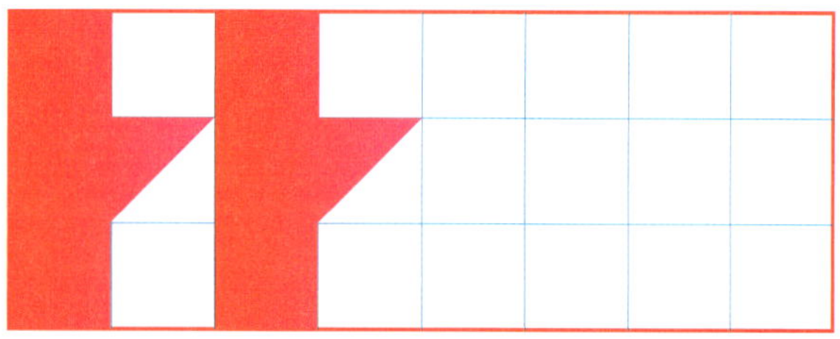

2. Finish this using pattern blocks.

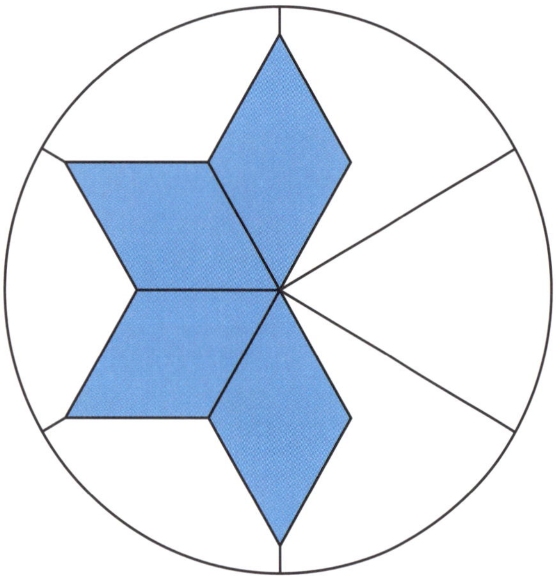

3. Flip the pattern across the line.

ICT

Use a computer drawing program like Microsoft® Word® to draw flips, slides and turns.

- Follow directions such as: 'Go to **Insert** → **Clip Art** (or **Shapes**)'.
- Use anchor points and copy and paste to flip, slide and turn.

Appendix 1: Australian Money

1 Write the value of each coin.

2 Match each banknote to the correct label.

$20
twenty dollars

$50
fifty dollars

$100
one hundred dollars

$10
ten dollars

$5
five dollars

3 Write the value of the banknotes in order from smallest to largest.

Appendix 2: Money

1 Write the total value of each group of banknotes.

a

b

c

d

2 Circle the larger amount.

a

b

c

3 Write the amounts in order from smallest to largest.

a

b

c

 ISBN 978 1 4886 2181 9

Progress Test 1A

Number and Algebra

1 Write as a numeral:

a 3 tens ☐ b 8 tens ☐

c 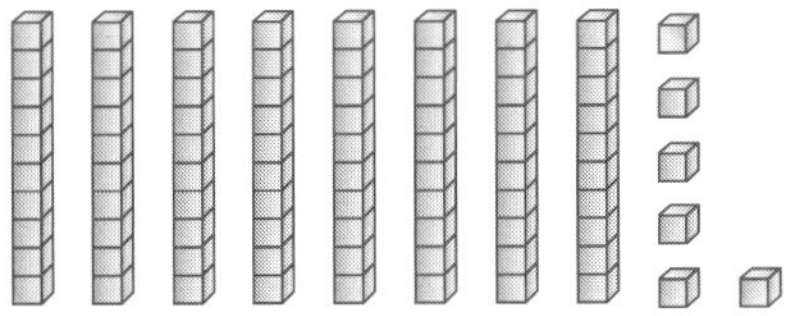 ☐

d 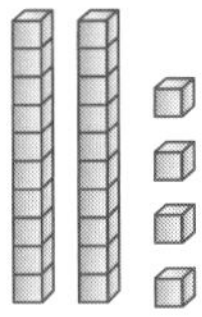 ☐

e 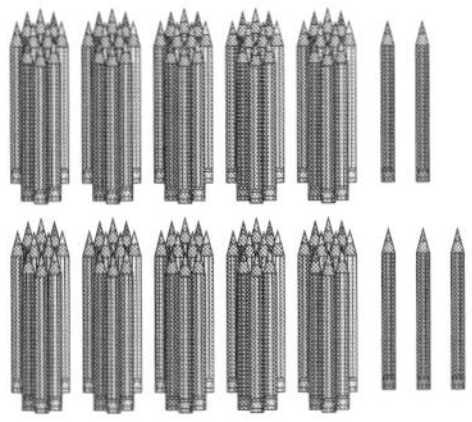 ☐

2 Complete each pattern.

94, 96, 98, ☐, ☐, ☐

74, 84, 94, ☐, ☐, ☐

15, 20, 25, ☐, ☐, ☐

80, 70, 60, ☐, ☐, ☐

3 Write the numbers **before** and **after**.

a ☐ 129 ☐

b ☐ 86 ☐

4 a $2 + 6 =$ ☐ b $3 + 7 =$ ☐ c $5 + 1 =$ ☐

d $9 + 4 =$ ☐ e $7 + 8 =$ ☐ f $10 + 10 =$ ☐

g $8 + 9 =$ ☐

5 a $10 - 6 =$ ☐ b $8 - 8 =$ ☐

6 a $6 + 6 =$ ☐ b $8 + 8 =$ ☐

$6 + 7 =$ ☐ $8 + 7 =$ ☐

$6 + 5 =$ ☐ $8 + 9 =$ ☐

7 

How many groups? ☐

How many in each group? ☐

How many altogether? ☐

☐ groups of ☐ = ☐

☐ + ☐ + ☐ = ☐

8 Colour:

a one quarter b one half

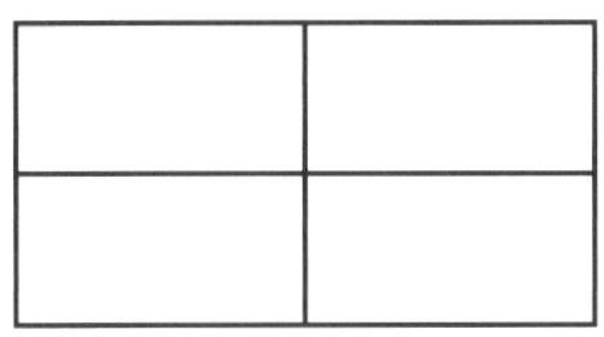 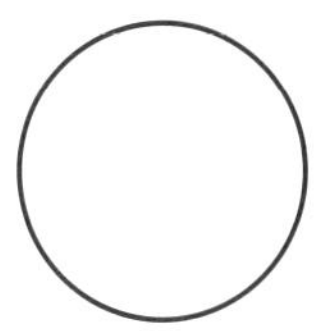

Progress Test 1B

Measurement

❶ Write the time.

a

☐ o'clock

b

☐ o'clock

c

d

e

f

g

h

❷ Show the time.

a

2 o'clock

b

half past 3

c

6 o'clock

d

half past 7

e

a quarter past 3

f

a quarter to 8

g
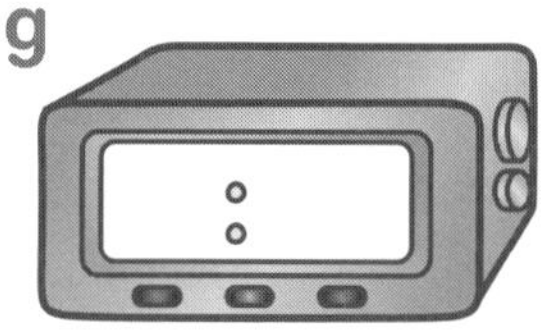

a quarter past 4

h

a quarter to 5

❸

February						
Sun	Mon	Tues	Wed	Thurs	Fri	Sat
			1	2	3	4
5	6	7	8	9	10	11
12	13	14	15	16	17	18
19	20	21	22	23	24	25
26	27	28				

a Colour February 22nd.

b Tick the second Monday.

❹

Jar	8 cups
Bottle	6 cups

a Which container has the smallest capacity? ☐

b How many more cups can the jar hold than the bottle? ☐

 ISBN 978 1 4886 2181 9

5 Use hand spans to measure this page.

Width of page: ☐

Height of page: ☐

6

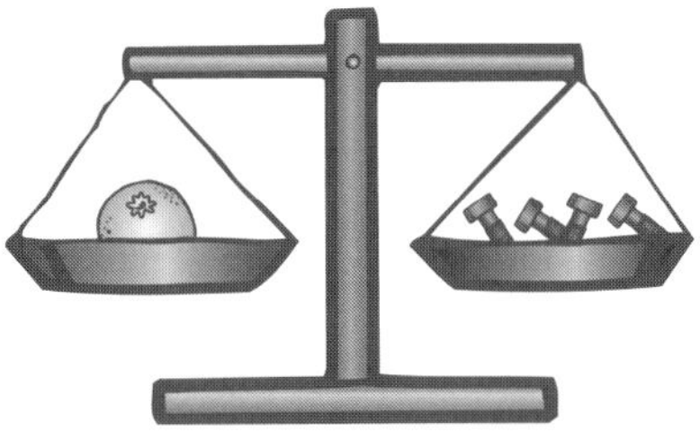

Mass = ☐ bolts

Geometry

7

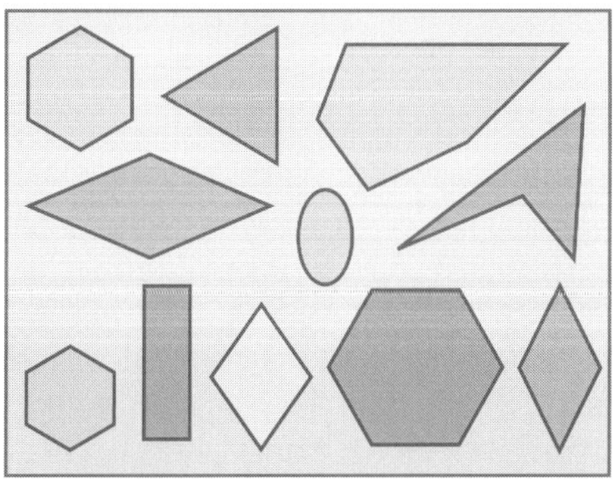

In this picture, how many of each shape are there?

a ovals ☐

b rhombuses ☐

c hexagons ☐

8 Circle each pair of parallel lines.

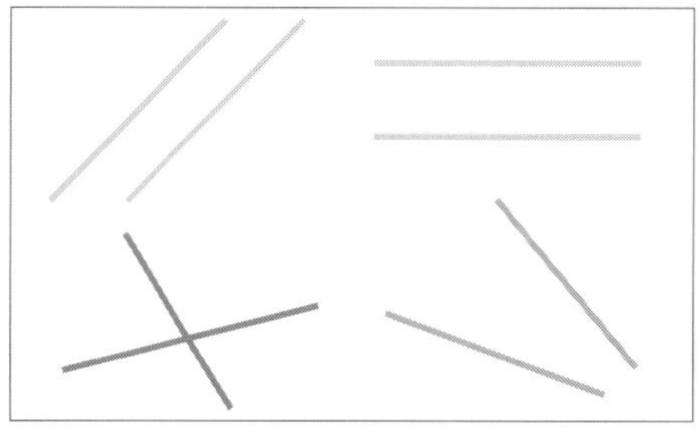

9 Draw lines to match.

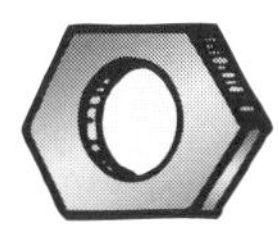

stacks

slides

rolls

no corners

flat surfaces

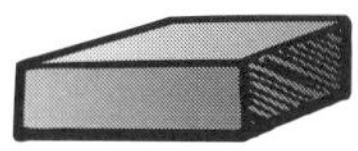

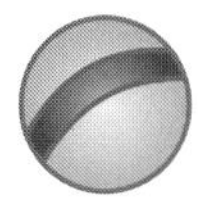

10

Circle the correct description for each.

a The pram is under/beside/above the girl.

b The car is in front of/above/between the house.

c The road is underneath/behind/in front of the girl.

 • • ISBN 978 1 4886 2181 9

Progress Test 2A

Number and Algebra

1 Complete the numeral expander, then write the number.

a

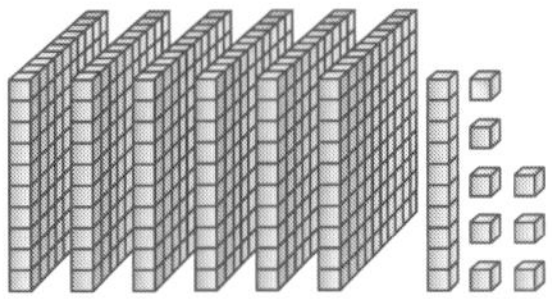

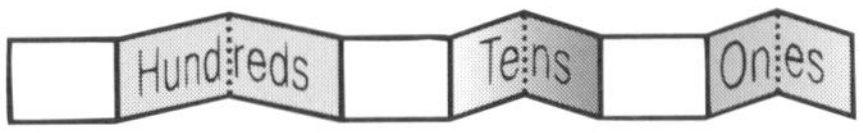

b

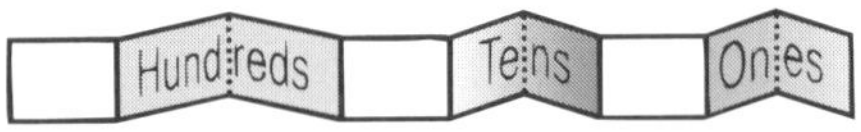

2 Write three different number sentences.

☐ + ☐ + ☐ = 19

☐ + ☐ + ☐ = 19

☐ + ☐ + ☐ = 19

3 Show 12 − 7 on the number line.

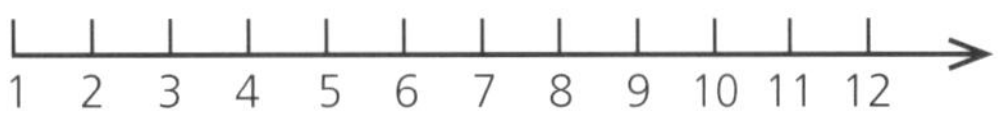

12 − 7 = ☐

4 **a** I had 12 pencils but I lost 5 of them. How many do I have left?

☐
− ☐
☐

b Dom has 8 stickers. Pam has 4. How many do they have altogether?

☐
\+ ☐
☐

5 Write number sentences to match the picture.

2 + 6 = ☐   8 − ☐ = ☐

6 + 2 = ☐   8 − ☐ = ☐

6

a How many stars? ☐

b How many rows? ☐

c How many stars in each row? ☐

7 **a** 3 rows of 2 = ☐

b 2 groups of 4 = ☐

c 4 × 2 = ☐

 • *AUSTRALIAN SIGNPOST MATHS 2* • ISBN 978 1 4886 2181 9

8 How many balls would be in this many cans?

a 2 cans

b 5 cans

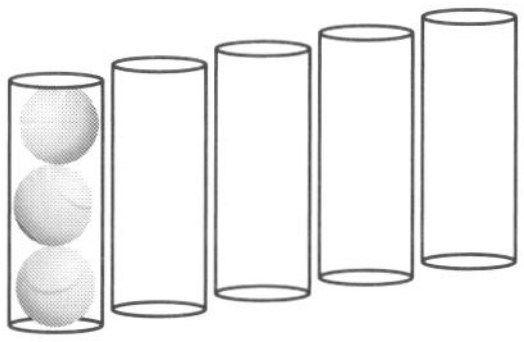

9 Share 8 counters between 2 groups.

One share =

10

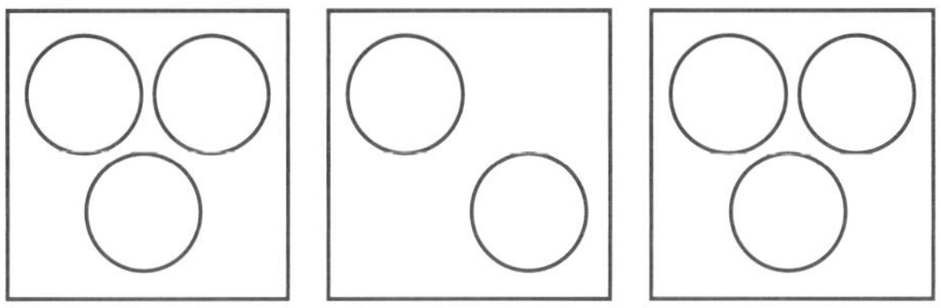

Are these fair shares?

11 What fraction is shaded, $\frac{1}{2}$ or $\frac{1}{4}$?

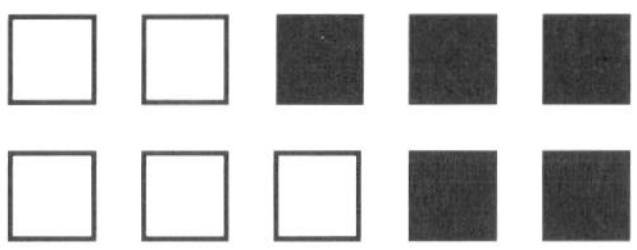

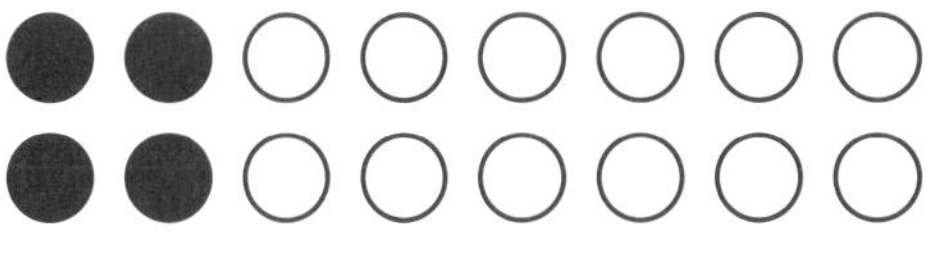

12 How many are in:

a half of this group?

frogs

b a quarter of this group?

counters

13 Write the fraction shaded.

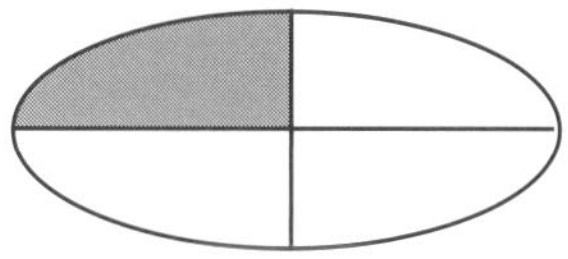

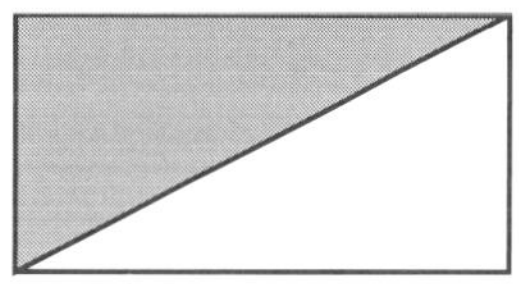

Measurement

1 Write the time shown.

a

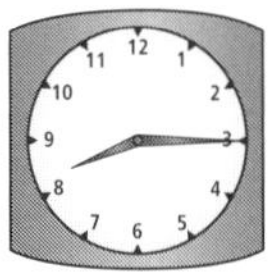

b

2 How many minutes in this much time?

a a quarter of an hour

b half an hour

c three quarters of an hour

3 Write two different ways of saying this time in words.

4

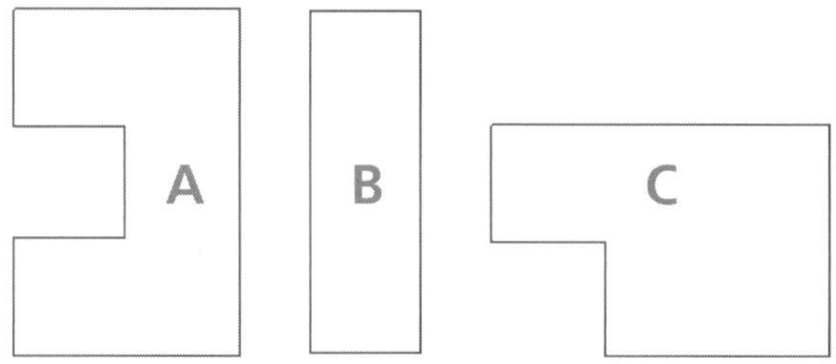

Which two shapes have the same area?

Geometry

5 Name each shape.

a

b

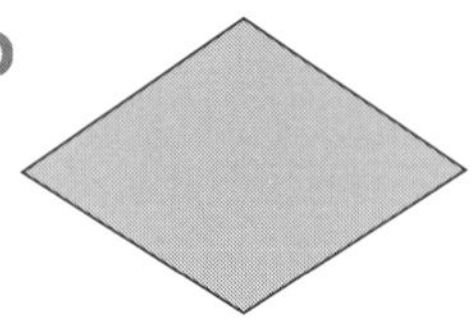

c

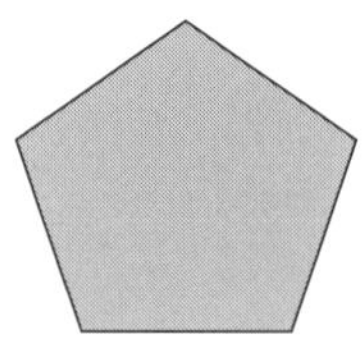

d

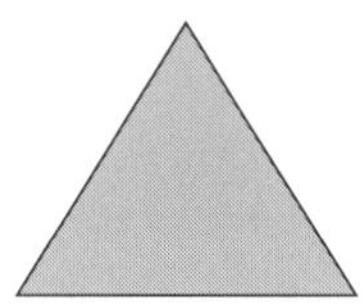

6 Name and describe each 3D object.

a

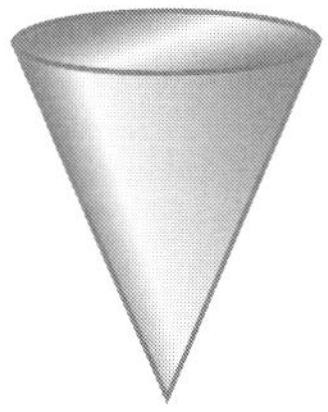

b

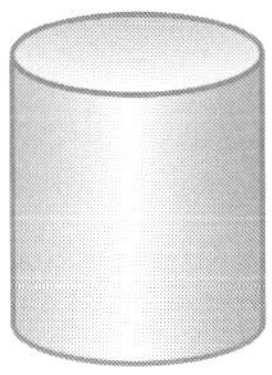

Location

7

a What is to the right of the bird?

b What is to the left of the bird?

Statistics

8 **Stickers given**

a How many stickers was Luke given?

b How many stickers were given altogether?

9 Use this list to complete the table and graph.

Favourite Colours:

blue, red, red, red, blue, red

Table:

Blue	
Red	

Graph:

Blue				
Red				

Probability

10 A

B

Which is more likely?

Progress Test 3A

Number and Algebra

❶ Write the number shown.

a 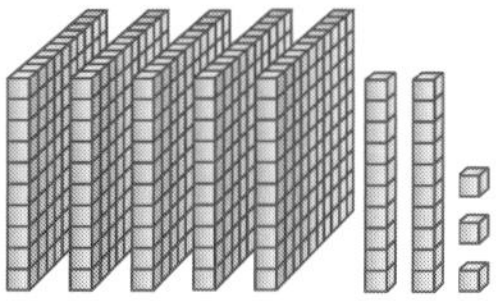 ☐

b 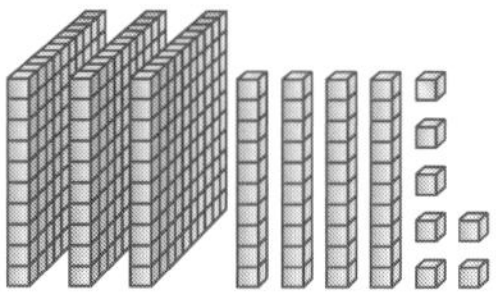 ☐

c 5 Hundreds 0 Tens 9 Ones ☐

❷ Write the **largest** number you can using the digits 3 and 7. ☐

❸ Write in order from **smallest** to **largest**.

109, 190, 119, 93

☐

❹ **a** 6 + 8 = ☐

b ☐ + 7 = 15

c 5 + ☐ = 13

d 14 – 8 = ☐

e 12 – 6 = ☐

f 7 less than 11 = ☐

g 5 + ☐ + ☐ = 10

h 12 + ☐ + ☐ = 20

❺ 17, 27, 37, ☐, ☐, ☐

98, 88, 78, ☐, ☐, ☐

54 + 20 = ☐

39 + 40 = ☐

75 – 10 = ☐

76 – 30 = ☐

❻

a	**b**	**c**
3	6	1
8	5	9
2	3	2
+ 4	+ 4	+ 8
☐	☐	☐

❼ Use bridging to tens.

48 + 5 = 48 + ☐ + ☐

= ☐ + ☐

= ☐

❽ 3 × 3 = ☐

2 × 3 = ☐

4 × 3 = ☐

9 Write the fraction shaded.

 ☐ 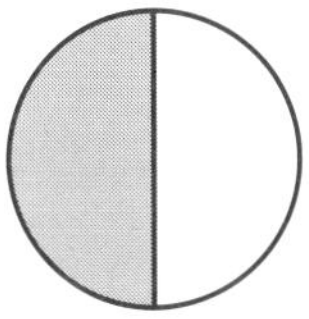 ☐

10 Colour half of each shape.

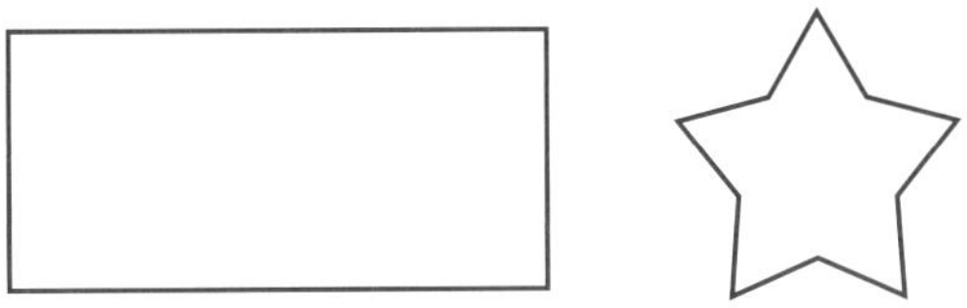

11 What part of each group is shaded?

a

☐

b

☐

Colour part of each group to to match the given fraction.

c $\frac{1}{4}$

d $\frac{1}{2}$

12 Colour the coins you would need to pay for each item.

a

b

13 Write the total value of the coins.

a

☐

b

☐

14 3, 8, 13, ☐, ☐, ☐

70, 65, 60, ☐, ☐, ☐

43, 45, 47, ☐, ☐, ☐

 • *AUSTRALIAN SIGNPOST MATHS 2* • ISBN 978 1 4886 2181 9

Progress Test 3B

Measurement

1 Write the time.

a

b

c

d

Show the time.

e

a quarter to 9

f

a quarter past 7

g

a quarter to 6

h

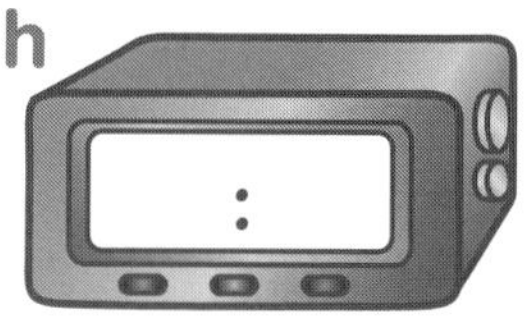

a quarter past 5

2 Write the month **after**:

a June

b October

3 a

Mass = ☐ bolts

b

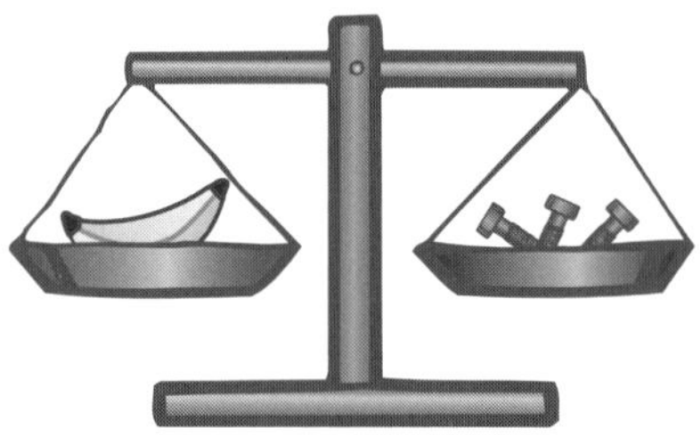

Mass = ☐ bolts

c Circle the correct answer:

The jar/banana is ☐ bolts heavier.

4 Which has the greater mass, a balloon or a coffee mug?

5

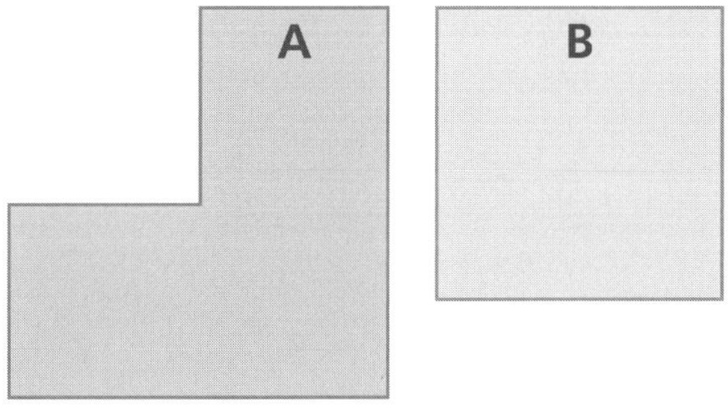

Which area is larger? ☐

Geometry

6 Name and describe each shape.

a 

b 

7 Is it a **flip** or a **slide**?

a 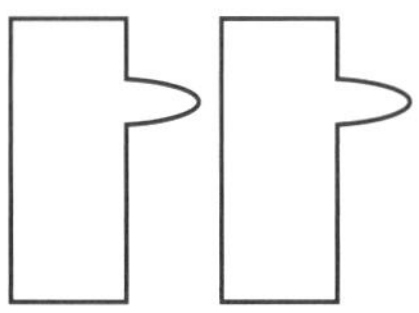

b 

8 Describe each 3D object.

a 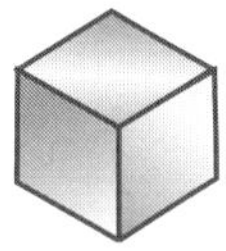

b 

9 Name each 3D object by choosing a name from this list: cone, pyramid, triangle, oval, cylinder, prism, rectangle, sphere.

a 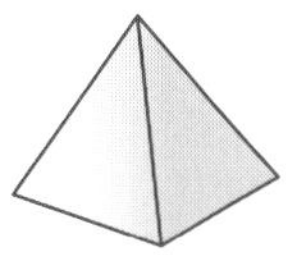

b 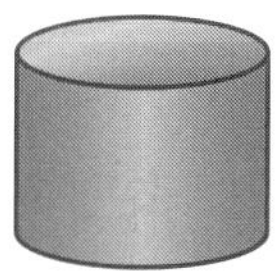

c 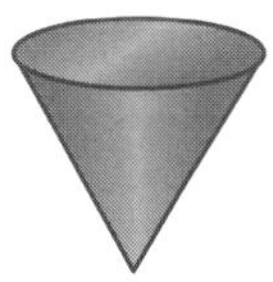

d 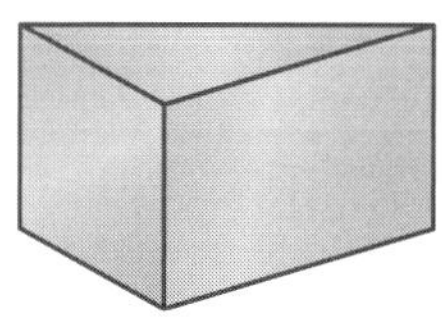

e How many faces does the object in part **d** have? ☐

Statistics

10 **The Weather on Our Holidays**

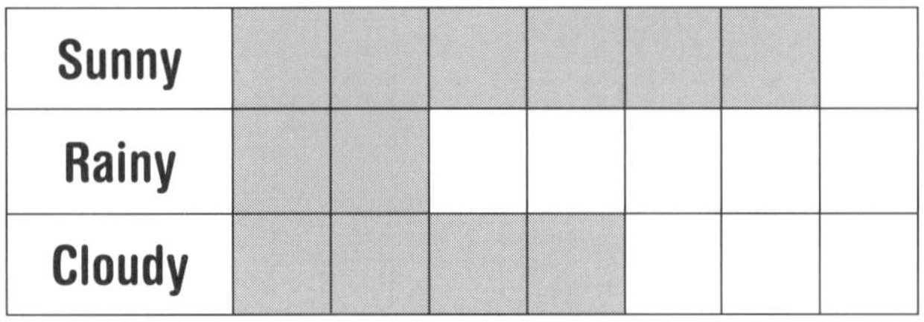

Each square stands for 1 day.

a How many days were sunny? ☐ days

b How many days were cloudy? ☐ days

c How long was our holiday? ☐ days

d Complete this table for the graph.

Sunny	
Rainy	
Cloudy	

11 Write the number shown.

a  ☐

b  ☐

c These are called ☐.

12 Use this table to complete the graph.

Favourite Place

Park	4
Shops	5
School	3

Favourite Place

Park	Shops	School

Probability

13 **A**

B

Which is more likely to be seen? ☐

Progress Test 4A

Number and Algebra

1 Circle the correct answer.

a 127 is less than/more than 120.

b 78 is less than/more than 90.

c 59 is less than/more than 82.

2 Round to the nearest ten.

a 57 ☐

b 34 ☐

3 Use the split strategy.

a 42 + 26

= ☐ + ☐

= ☐

b 87 – 35

= ☐ + ☐

= ☐

4 Use the jump strategy.

a 47 + 24 = ☐

b 82 – 13 = ☐

5

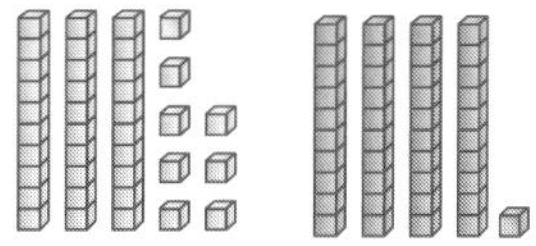

38 + 41 = ☐

6

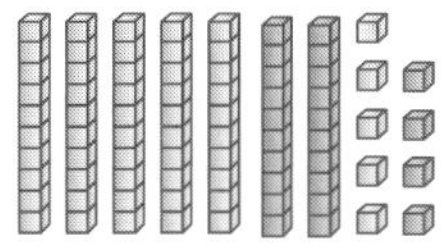

79 – 24 = ☐

7

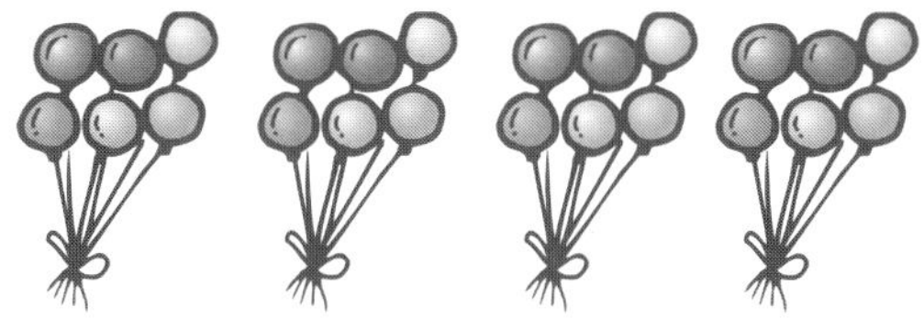

a 1 × 6 = ☐

b 2 × 6 = ☐

c 3 × 6 = ☐

d 4 × 6 = ☐

8

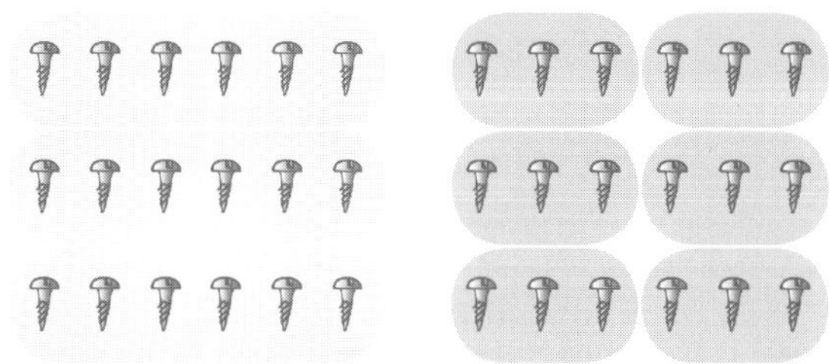

a How many are in 3 groups of 6? ☐

b How many are in 6 groups of 3? ☐

c True or false? 3 × 6 = 6 × 3 ☐

 • *AUSTRALIAN SIGNPOST MATHS 2* • ISBN 978 1 4886 2181 9

9 a $7 \times 2 =$ ☐

b $3 \times 5 =$ ☐

c $4 \times 10 =$ ☐

10

a 20 shared by 4 = ☐ each

b $20 \div 4 =$ ☐

c How many 10s in 20? ☐

d $20 \div 10 =$ ☐

11

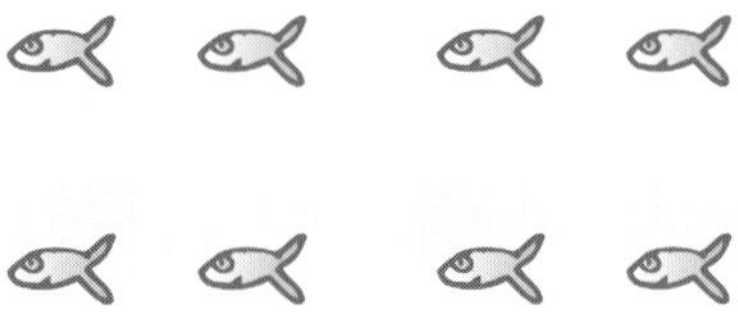

a How many groups of 2 fish? ☐

b $8 \div 2 =$ ☐

c How many groups of 4 fish? ☐

d $8 \div 4 =$ ☐

12 How many students could be given:

a 2 balls? ☐

b 3 balls? ☐

13 a Circle $\frac{1}{2}$ of the vases.

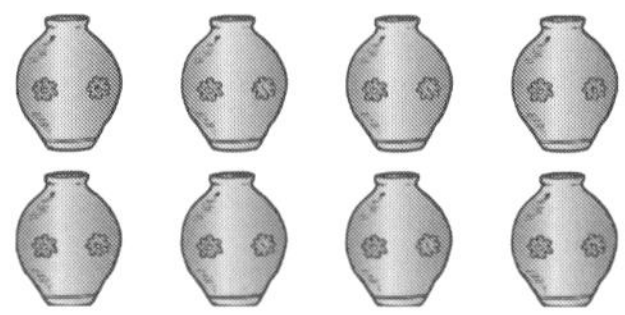

What part is shaded?

b

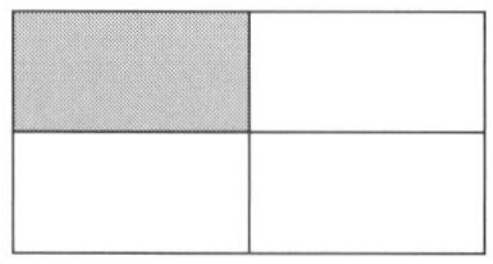

☐

c

☐

14

a Fraction shaded =

☐ out of ☐

b Is the part shaded **less** than one half? ☐

 ISBN 978 1 4886 2181 9

⑮ What is the total value?

a

b

⑯ Colour the coins you would need to buy each item.

a

b

Progress Test 4B

Measurement

❶ January, February, ☐,
April, ☐, ☐,
☐, ☐,
September, ☐,
November, December

❷ a The season after winter is ☐.
b The season before winter is ☐.
c The first month of spring is ☐.
d The first month of summer is ☐.

❸ How many days are in each month?
a March ☐
b September ☐
c December ☐

❹ Would you use **minutes**, **hours**, **days** or **months** to measure the time to:
a draw a picture? ☐
b paint a house? ☐

❺ Answer these questions for the calendar below.

March						
Sun	Mon	Tues	Wed	Thurs	Fri	Sat
		1	2	3	4	5
6	7	8	9	10	11	12
13	14	15	16	17	18	19
20	21	22	23	24	25	26
27	28	29	30	31		

a On what day does March begin? ☐
b How many Tuesdays in March? ☐
c What is the date of the 2nd Friday in March? ☐
d What day is the 15th of March? ☐

❻ Which has the smaller area, a newspaper page (**A**) or this page (**B**)? ☐

Geometry

❼ Draw each shape.

rectangle

parallelogram

rhombus

trapezium

❽ Is it a **flip**, **slide** or **turn**?

a

b

❾ Name each 3D object.

a

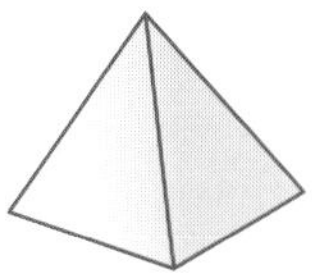

b

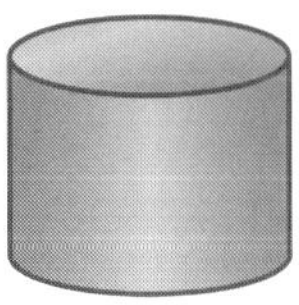

c

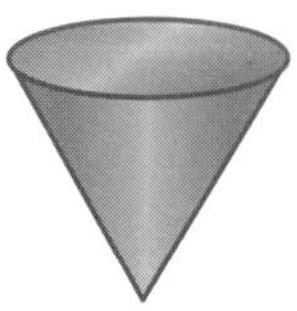

d

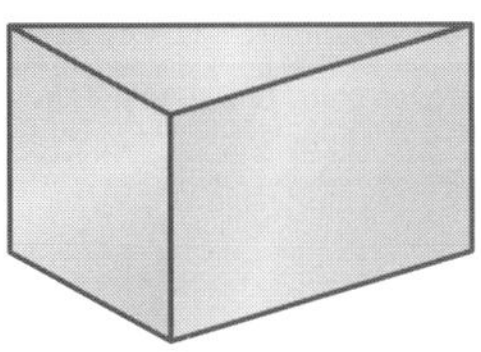

Location

❿ Start at ×. Move 4 right, 2 up, 7 left and 1 down. Colour the square where you finish.

				×					

⓫ Write directions to get from **A** to **B**.

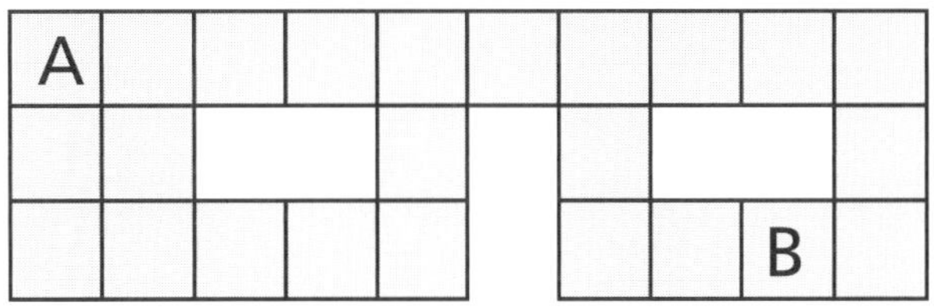

Colour the shortest route from **A** to **B**.

Statistics

12 Make a tally to show the different Christmas baubles.

Christmas Baubles	
Green	
Red	
Yellow	

13 Use this list to complete the table and graph.
Number of Sisters:

0, 1, 1, 3, 0, 1, 1, 2, 2, 0

Table:

0	
1	
2	
3	

Graph:

0	**1**	**2**	**3**

Probability

14 **a** List all the possible outcomes for this spinner.

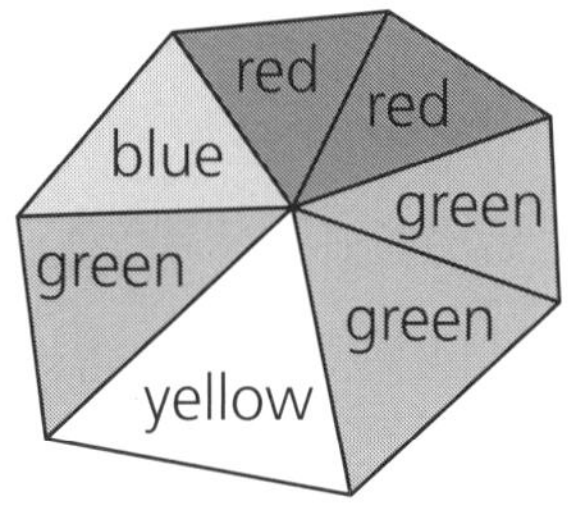

b Which colour is most likely to be spun?

15 A ball is taken from this jar.
Is it **certain**, **likely**, **unlikely** or **impossible** that:

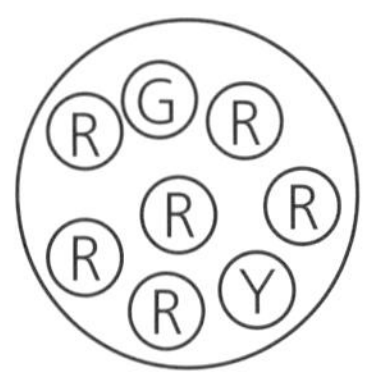

a an **R** could be taken?

b a **B** could be taken?

c a **G** could be taken?

• *AUSTRALIAN SIGNPOST MATHS 2* • ISBN 978 1 4886 2181 9